# Write Your Own Proofs

## in Set Theory and Discrete Mathematics

Amy Babich and Laura Person

ZINKA PRESS

Grateful acknowledgment is made for permission to reprint excerpts from the following.

THE BEST OF MYLES, by Flann O'Brien, © 1968. Reprinted by permission of The Dalkey Archive Press, Illinois State University.

Excerpt from THE BLUE FLOWER by Penelope Fitzgerald. Copyright © 1995 by Penelope Fitzgerald. Reprinted by permission of Houghton Mifflin Company. All rights reserved.

From CHERRY by Mary Karr, copyright © 2000 by Mary Karr. Used by permission of Viking Penguin, a division of Penguin Group (USA) Inc.

"The Kid Who Learned About Math on the Street" from THE FOUR ELEMENTS, by Roz Chast, ©1988 by Roz Chast, reprinted with the permission of the Wylie Agency

LITTLEWOOD'S MISCELLANY, by J. E. Littlewood, edited by Béla Bollobás, Copyright ©1986 by B. Bollobás, reprinted by permission of The Cambridge University Press.

JR, by William Gaddis, ©1951 by William Gaddis, reprinted with the permission of the Wylie Agency.

MURPHY, by Samuel Beckett. Copyright © 1938 by Samuel Beckett. Reprinted by permission of Grove/Atlantic Inc.

NIAGARA FALLS ALL OVER AGAIN, by Elizabeth McCracken. Copyright © 2001 by Elizabeth McCracken. Reprinted by permission of Random House, Inc.

From ON BEYOND ZEBRA!, by Dr. Seuss, Copyright TM & copyright © by Dr. Seuss Enterprises, L.P. 1955, renewed 1983. Used by permission of Random House Children's Books, a division of Random House, Inc.

THE THIRD POLICEMAN. by Flann O'Brien, © 1967. Reprinted by permission of The Dalkey Archive Press, Illinois State University.



Library of Congress Control Number: 2005906010
ISBN-13: 978-0-9647171-7-6
ISBN-10: 0-9647171-7-4

PRINTED IN THE UNITED STATES OF AMERICA
AT MORGAN PRINTING IN AUSTIN, TEXAS

In memoriam

David Spellman
(1954 – 1997)

# Acknowledgments

We would like to thank Shwu-Yen T. Lin and You-Feng Lin for permission to use the first half of their book, *Set Theory with Applications* (1985, Book Publishers, Tampa, Florida) as the basic template for this book. We have taken our basic outline, many problems, and some examples and explanations from Lin and Lin.

The authors would also like thank the following people for their input and feedback: Adam, Cheri, Jo and Walter Babich, Vasily Cateforis, Orlando Canelones, Sandra Cox, Harold Ellingsen, Warren Hamill, Lynn Huang, Victoria Klawitter, Brook Landor, Sam Moczygemba, Vivek Narayanan, Jeff Reeder, Irene Schensted, Kang-Ho Song, and Armand Spencer (author of Dr. Spencer's Mantra).

Special thanks go to Mike Librik for typesetting this book, and for designing and producing the front and back covers. The typesetter would like to thank MacKichan Software for their fast, free, and industrious support of their software *Scientific Word 5.0*, with which this book was typeset.

# Contents

# Foreword A
# This Book and How to Use It

About twenty years ago, American universities began to offer a mathematics course (called variously Set Theory and Logic or Discrete Mathematics, at different institutions) designed to serve as an introduction to mathematical proof. The usual assigned textbook for this course is either Chapter Zero of a graduate mathematics text, or one of the recent textbooks on Discrete Mathematics (by Grimaldi, Epp, Rosen, et al.). Neither choice of textbook is ideal for the novice professor assigned to teach this course. Chapter Zero of a graduate textbook is mere outline, with very few exercises, and the Discrete Mathematics books are all much too large for a one-semester course. Nearly always, the rookie instructor for this course must make her/his own class notes and supplementary problem sets. This takes time and effort, and the first-time instructor is often short of both.

Like other instructors of this course, we made our own class notes. We also traded our notes back and forth, making use of each other's ideas. We used the first half of *Introduction to Set Theory and Logic* by Lin and Lin, as a model, but we changed it to suit our purposes. We did not find that the section on deductive reasoning (featuring syllogisms, Modus Ponens, and Modus Tollens) helped our students to write proofs, so we jettisoned it. There is a great deal of material in this sort of course, and we move through it slowly, at the students' pace. We omit whatever we don't find helpful.

We used these class notes for several semesters, changing them as we went along. The main change that occurred with time and use was the addition of examples and exercises in sections that the students found difficult. The notes, even in their most rudimentary form, were noticeably more effective than our old textbooks for helping students to learn. Students said that they liked the notes, but wished that they were a real book.

So we set about turning the notes into a book. (This turned out to be a lot of trouble.) We tried to write down the explanations we gave orally in class. These appear as "Remarks" in the text. The Remarks are there to help whoever may find them helpful. If a Remark seems unhelpful or confusing, feel free to ignore it. The Remarks are optional aids to understanding.

We have been at some pains *not* to include all possible material in this text. When writing a book of this sort, one quickly comes to understand why the available books (by Epp, Grimaldi, Rosen, et al.) are so huge. The material here has very natural connections to number theory, probability, graph theory, complexity theory, transfinite arithmetic, and so on. But we do not wish to include an introductory course in each of these subjects, although we acknowledge the strong temptation to do so.

We have tried to include slightly more than enough material for a one-semester course in Set Theory and Logic or Discrete Mathematics. The focus is on teaching the student to prove theorems and to write down mathematical proofs so that other people can read them.

Proving theorems takes a lot of practice. This text is designed to give the student plenty of practice in writing proofs. In our own classrooms, we usually have the students write up the homework problems on the blackboard. This slows down the class to the students' level of understanding. In the early chapters (particularly Chapters 1 and 2) we often have students write a few proofs during class. We do this in order to forestall the statement that students didn't do the homework because they didn't know where to start. "Where to start" in writing a proof is at least half the content of this course. When students know where to start, they often write proofs correctly. A false start is the most common mistake.

Proving theorems used to be taught on a sink-or-swim basis. Those who swam were those with good mathematical habits — students who broke difficult propositions into pieces, played with new notation in order to learn it, constructed examples for themselves, and so on. Most students do not already have these habits. In our text we do some of this work for the student. We break theorems into pieces, and walk the student through examples and exercises. It is important to make sure that students get their hands dirty, use notation and write proofs themselves. Many students don't want to do this at first. They want to read, not write. They don't want to risk making mistakes. It's important to get them past this reluctance.

The material covered consists of propositional logic, set notation, basic set theory proofs, relations, functions, induction, countability, and some combinatorics, including a very small amount of probability. In the last few chapters we give the student some practice in constructing functions on sets of sets and sets of functions. (Every new level of abstraction is confusing for beginners and must be practiced.) Working with such functions gets the student used to them. This whole course is really an exercise in becoming accustomed to working with mathematical notation and writing proofs based on definitions. For this reason, the instructor should not feel compelled to cover as much material as possible. The most important thing is that the students start to get comfortable with reading and writing mathematical proofs. In our own classrooms we have usually covered only four or five chapters in a semester, not all seven. But a fast-paced class of students with good backgrounds might cover all seven chapters.

We do not assume any knowledge of calculus on the part of the student. At some universities, successful completion of a calculus course is a prerequisite for entry into this class. This seems unnecessary. In the usual textbooks for this course, only two or three limits are computed in the course of the semester. We have omitted these limits, as we would like this course to be accessible to people who have not studied calculus.

We hope that you will find this book as effective and enjoyable to use as we have found it in trial versions.

# Foreword B
# Words and Numbers: Mathematics, Writing, and the Two Cultures

A newspaper columnist recently wrote a rather facile article which divided educated human beings into "word people" and "number people." This is more or less the division that C. P. Snow called "the two cultures." On the one hand, we have language, literature, history, and philosophy; on the other hand, science and engineering. It is not generally recognized that mathematics sits squarely in the middle of this rather artificial division. Part of mathematics is computational, and involves manipulating numbers. And a very important part of mathematics, largely ignored by our pre-college educational system, involves words, reasoning, and proofs based on verbal definitions. This book is concerned primarily with the verbal side of mathematics.

There is reason to believe that people who study language, literature, philosophy, and suchlike things will be interested in the verbal side of mathematics. There is reason to expect that many "word people" actually have considerable talent for verbal mathematics. Mathematics has a history of significant contributions from "amateurs" — people who do something non-mathematical for a living. Leibniz was a diplomat; Fermat was a lawyer. Descartes, Leibniz, and Pascal are as famous for their philosophical writings as for their mathematics. The division of intellectuals into "two cultures" — the verbal and the numerical — is a twentieth-century phenomenon, and an unnecessary one. "Word people" have always made enormous contributions to mathematics. Unfortunately, under our current educational system, "word people" are unlikely ever to encounter the side of mathematics which will speak to them.

In the United States, the numerical/verbal, science/humanities division is also seen as a masculine/feminine division. Most women in the U.S. are classified as being on the verbal side. This is not an accident; it is, at least in part, the result of years of propaganda. "Girls can't do math," used to be a mantra in schools. When a girl did very well in mathematics, she was often told that, as she grew up and became womanly, she would be surpassed by boys in mathematics.

It is a very bad idea, in general, to promote the notion that women just can't do certain things. Since it is still primarily women who educate children, the result of such propaganda is that eventually neither men nor women are able to do the things in question. We used to hear, "Girls can't do math." Now we hear, "Americans can't do math."

There is another U.S. custom which has the unintended effect of frustrating the "verbal" student with an interest in mathematics. This is the convention that university mathematics education nearly always begins with the standard U.S. calculus class. The standard U.S. calculus class is not a proof class. The professor usually proves some theorems, and there are always proofs in the calculus book, but writing proofs is not the focus of the course. Since one cannot really understand proofs until one can write proofs correctly oneself, the more "verbal" student winds up feeling unsatisfied by the calculus class. There are many "verbal" students who take calculus, make a good grade, and then never look at mathematics again. They want something that the standard U.S. calculus class does not give them, although they are probably unable to say exactly what it is that they want. What they want, we believe, is an approach to mathematics through words, through definitions and proofs, through writing.

Thus, for "verbal" people, it makes sense to begin the study of mathematics with an introduction to the language and conventions of proof. A student who has done the exercises in this book will be in a good position to study number theory, abstract algebra, or real analysis. A "verbal" student will probably be happier studying real analysis before calculus, rather than the other way around.

It is usual in U.S. mathematics departments to classify calculus as a freshman course and all proof classes as "advanced." This is primarily because the U.S. mathematics curriculum is designed for engineers and scientists. Calculus is taught early so that the student may start learning physics as soon as possible. Since many engineering and science students, in contrast to "verbal" students, are more comfortable with computations than with language, proof classes are relegated to junior or senior year. Moreover, successful completion of a calculus class is often a prerequisite for entry into a proof class. This requirement ignores the needs of the verbal student, because the verbal student is assumed to be uninterested in mathematics.

The authors of this book have taught this introductory proof class several times. We have found that the students who do best in the course are those who are most sensitive to language.

We would like to see an end to "the two cultures." We would like the language and conventions of mathematics to be part of the intellectual birthright of all students. We would like to stop reading in the newspaper that "Americans can't do math." And we would like to see a return to the notion that amateurs — novelists, diplomats, classicists, historians, lawyers, librarians and so forth — not only can enjoy mathematics, but can contribute significantly to it.

Mathematics belongs essentially not only to the numerical, but also to the verbal culture. Mathematics is free, and should belong to us all, not just to a privileged elite. We who speak the language of mathematics have no interest in being a privileged elite. One thing we like about mathematical notation is that it makes communication possible between mathematicians who have no other common language. Mathematical language is for letting people in, not for keeping people out. To learn mathematics is to love it. We would like more people to share this experience.

For this reason, the authors of this book do not assume that the reader has studied calculus. We think that there are many people (especially verbal people) who can enjoy calculus only after completing an introductory class like this one and a first course in real analysis. In the U.S. at the present time, such people generally stop studying mathematics either before or just after calculus, and never see a proof class at all. We hope that this situation will change. Currently, it often turns out that only a few undergraduate mathematics students like or do well in proof classes. It may be that some of the best proof students — the verbal students — are inadvertently being excluded. It might even turn out that girls and Americans are good at mathematics after all.

# Foreword C
# Mathematical Proof as a Form of Writing

Lucidity is nine-tenths of style. *Elisha said to the boys: If you do that again I will tell a big bear to come and eat you up. And they did. And he did. And it did. (It could do with the odd tenth.)*

*J.E. Littlewood,* Littlewood's Miscellany

This book is about the experience of writing and reading mathematical proofs. A mathematical proof is a peculiar and amusing sort of written document. The language is formal. It sounds solemn and grandiose and absurd. It can also sound very elegant.

Of course, the way the proof looks or sounds is not the point. The point of the proof is the theorem it proves and the way it proves the theorem. The shape of the logical argument is what makes a proof elegant.

Mathematical writing tries to be as clear as possible. This is much harder than it might seem. Our language of sets, quantifiers, relations and functions is a great help in writing proofs and in understanding one another's proofs.

Since the goal of a mathematical proof is to show as clearly as possible how a piece of deductive reasoning works, it is not a flaw in mathematical writing to use the same sentence structure several times in a row.

Variation in language just for its own sake should, in general, be avoided, especially by beginners. (Of course, beginners love to play with language. This isn't really a bad thing, and, even if it were, could not be helped. But be warned that some professors may find it more annoying than amusing when you give all your variables funny names and salt your proofs with such phrases as *mutatis mutandis* or *per impossible*.) The focus should be on the argument.

In some ways, mathematical writing is like poetry. A mathematician, like a poet, gets stuck and requires inspiration. Of course, it does no good to wait around for inspiration to strike; the only thing to do is to attempt to write the proof or the poem, or at least go through the motions. Here the mathematician may have the advantage over the poet that there are several known strategies to try. (But then, the poet also has a bag of tricks.) Often known strategies don't seem to work. Then the mathematician or poet goes out for a walk (or even just as far as the next room) and an idea starts to form. The person has an idea, but doesn't yet know what the idea is. The hope is to put the idea into language that will clearly reveal its lineaments to the writer as well as to readers.

In other ways, mathematical proofs are like plays. They are rather formal plays, in which each character must be introduced before it (mathematical objects seem genderless) can play its role in the drama.

As readers of fiction and poetry we often wish that writers in general knew more about mathematics. For us, mathematics is a part of life. But characters in books (especially female characters) usually seem untouched by it. As women, we are grieved when female writers we like seem to feel that mathematics is masculine, or boring, or "linear." Mathematics is none of these things.

While we think it inadvisable for the beginner to write proofs in the style, say, of S.J. Perelman, we would like to have read Perelman's mathematical pastiches, had he written any. We think that if more writers knew more verbal mathematics, some entertaining books might be written.

Writers of novels, poems, histories, and so on may find that they enjoy writing mathematical proofs. Writers may find that their mathematics and their non-mathematical writing enrich each other. But there are some people who hate to write, in some cases because their native language is not English and they make mistakes, forgetting articles, misspelling words, mixing up singular and plural. For these people, we have good news.

In many ways, mathematical English is much easier to write than conversational, journalistic, or literary English. Mathematical English uses a limited vocabulary. We introduce variables and write formally and redundantly. This means that a mathematician who knows no English may very well understand a talk in mathematical English, and even learn a few English phrases in the course of it. (It would be hard, for example, to avoid learning the phrase "such that.") It is very common for people to write mathematical papers in a language (often English) that is not their own. Therefore, nobody cares very much about small grammatical solecisms such as misuse of articles or omission of plural endings. All anybody really cares about is understanding the proof. If the specifically mathematical parts of the language are correct, the proof will be understood.

Mathematical language is redundant. We keep reminding ourselves what sets our characters belong to. This is useful, because written mathematical documents nearly always contain errors. Mathematicians like to get things right, and try hard to produce error-free documents. However, this is nearly impossible. Despite our efforts, there will be a missing subscript, or "$=$" with be printed instead of "$\neq$." The redundancy of mathematical language helps the reader to catch and correct errors.

Occasionally, in our experience, a student shows up in class with a laptop computer and proposes to take notes and do the homework on the computer. This turned out well only in one case, where, anomalously, the student was blind. As far as we know, anyone who is able to write (or print) by hand will be most successful writing proofs by hand. Probably one reason for this is that, when writing proofs by hand, we don't just write but also fool around on the side, calculating, drawing pictures, writing down thoughts and rejecting most of them. The computer is not so well set up for this. Also, it is by no means as easy to type mathematics as to type ordinary language. Furthermore, it can be better not to have your thinking tied to a machine that cannot be with you everywhere. It can be fun to carry an unsolved problem in your head. If you can work by hand on a scrap of paper, it's easier to play with the problem in times of boredom or distress. Also, writing by hand seems somehow to make a stronger impression on the mind than typing.

Some students will find it hard to be patient enough to write things down. (This is particularly true of high school students, who like to rush through problems.) We would suggest that students cultivate a taste for the slow, deliberate writing of proofs. A person in too big a hurry will not be able to learn to prove theorems. It can actually be fun for a fast-paced person to slow down for a change and look at the scenery.

It may seem that, so long as the argument presented is correct, there is no difference between a badly-written proof and a well-written one. This is not true at all. A well-written proof is easy to follow. (This is an odd statement, because of course the argument presented may be inherently so difficult that no elucidation of it can really be "easy." The often-stated ideal of mathematical lucidity is that any uneducated person who reads the proof should understand it immediately. This is an exaggeration, but it has a point.) A badly written proof creates unnecessary difficulties.

Beginning students often regard all their mathematical textbooks as badly written, because they find the proofs hard to read. Actually, some of these textbooks are written very well indeed, but there is an inherent difficulty in learning to use mathematical language. When you write proofs yourself, you will appreciate the difficulties of writing good mathematical English. The less mathematics your intended audience knows, the harder it is to write mathematics clearly. It's easiest to write for experts.

For the student who hates writing but is truly interested in mathematics, we would suggest bearing in mind a remark attributed to Einstein, saying that he thought that his job at the patent office, which consisted of writing a short but complete and accurate description of each invention submitted for patenting, helped develop his mind in ways that were helpful for thinking of and writing down his physics. It's not enough to think of something; you also need to be able to describe it clearly to other people. Thus, it is worth learning to write mathematically in order to present one's own ideas in physics, engineering, or computer science. And it really doesn't matter if you can't spell.

In this course, the more "verbal" students will have an advantage. There are also people with deep mathematical insight who are "non-verbal." These people will struggle, but will be helped greatly in communicating with others about their ideas. Writers and students of writing will find the precision of mathematical language useful and its peculiar syntax entertaining.

We hope that every student who uses this book will enjoy the experience, but perhaps this is not possible. Should you truly dislike it, we would suggest stopping, and finding another approach to the subject. Mathematics should bring pleasure, not pain.

# Foreword D
# Amazing Secrets of Professional Mathematicians REVEALED!!!

Throughout our years of compulsory education, we study a subject called mathematics. We learn to count and do calculations in arithmetic, algebra, geometry, and calculus. (If we're lucky, we might get some proofs in geometry.) By the end of high school, students have some beliefs about mathematics.

Many high school students believe that talent in mathematics consists in quickness in calculating, preferably in one's head. The smartest people, it is believed, never write anything down and never study. High school students also tend to believe that success in mathematics is a special gift, a savant's magic facility for calculating, and that people who have this gift are essentially different from those who lack it.

In the world of verbal mathematics, these assumptions are stood on their heads. In verbal mathematics, in the realm of proof, we do not try to think as fast as possible. Instead, we take steps to *deliberately slow down* our thinking, so that we can more or less watch ourselves think. One way to do this is to write our thoughts down on paper. It is much easier to think mathematically if one writes things down.

When mathematicians speak of "reading" a mathematical book or paper, we do not mean "reading" in its ordinary sense. In fact, most mathematicians, when reading a mathematical document, actually copy out much of the text with pencil onto paper, word for word. As we do this, we frequently have to pause and prove to ourselves on paper that each claim made in the text is true. This is much harder work than the ordinary reading of a newspaper, textbook, or novel. It is akin to reading a novel in a foreign language which we have studied a little bit but don't know very well.

Ordinary reading, in which one simply sits and stares at the page while mentally hearing the words, is woefully inadequate for learning mathematics. Reading while underlining or highlighting is likewise useless. Ordinary reading, with or without highlighting, simply does not make a strong enough impression on the mind. "Reading mathematics" requires more attention than ordinary reading. The act of writing focuses the mind's attention. It helps to engage the part of the unconscious mind that is interested in mathematics.

Students who are new to verbal mathematics sometimes find it frustrating. Often such students are used to doing well in mathematics classes without effort. Suddenly effort is required, effort of an unaccustomed kind. And you, the good student who always makes 100, keep getting your homework back all marked up with corrections. It's as if, suddenly, you aren't good at math anymore.

Relax. Verbal mathematics is considerably harder than calculation. In this introductory course you will learn how to prove theorems using correct mathematical notation. To prove a theorem is to write a mathematical proof. A mathematical proof is written in the form of a paragraph or a few paragraphs. A proof is somewhat like a joke or a magic trick, in that it has a punchline. And, in general, a mathematical proof is written in formal mathematical language.

In this first course, you are learning the basic grammar of mathematical language. When you do your homework, you write proofs. Since you are writing essays in an unfamiliar foreign language, it's not surprising that you make many mistakes. As with a foreign language, you must make mistakes in order to learn how to write mathematics properly. That you can't instantly write the language perfectly does not mean that you are not good at math. Don't be afraid to make some beginner's mistakes. There's no other way to learn a foreign language.

Unfamiliar notation seems repellent and hard to use, you say. Is it really necessary to use quantifiers? Yes, it is necessary. Experienced mathematicians do not enjoy learning unfamiliar notation any more than you do. And yet we must put up with learning unfamiliar notation all the time. We put up with this annoyance because it's worth it to us. Unfamiliar notation is like vocabulary in a foreign language: wonderful to know but a nuisance to learn.

In a way, nobody is good at verbal mathematics. People don't learn to prove theorems instantly. It takes a few semesters of study to get good at it. Some people learn faster than others, but even the fastest don't learn everything perfectly the first time. Most professional mathematicians are not lightning calculators or savants. Their minds are like yours. What makes them "good at mathematics" is a strong interest in mathematics and years of experience in working with it.

Here are a few secrets known to every professional mathematician and graduate student, and unknown to nearly every beginner in verbal mathematics. (That these are secrets is not intentional; it has simply never occurred to anyone to state these truths in print.)

1. You need to learn the definitions by heart. Memorize the definitions. Make sure that you can write down the definitions from memory, using correct notation. Learn the definitions right away, not just before an exam. You shouldn't have to keep looking up definitions while writing your homework proofs.

2. Sometimes unfamiliar notation is really hard to understand. Suppose that you are reading a homework problem; that is, you have copied out on paper the statement of the problem. And suppose that the statement of the problem makes no sense to you whatever.

Any graduate student will tell you that this happens all the time: you want to work on a problem, but you don't quite even understand what the problem says. What you should do is at least write out the statement of the problem. (If you have already done this, do it again. You're trying to make an impression on your mind.) And if you have any ideas (even patently silly ones), play with them a little. Then stop working on it and do something else. The next day, after you have slept, it is very likely that you will understand the problem at least a little bit better. Your unconscious mind does some of the work while you sleep. (This is one reason why you should always start your homework as soon as possible. At least acquaint yourself with the problems, even if you don't solve them right away.)

**3.** You cannot learn to prove a theorem by watching a professor prove it on the blackboard, even if you take notes. Unless the professor makes a mistake (and this does happen occasionally), any proof done by a professor on a blackboard will look easy and seem to make perfect sense. (The professor, like a professional magician, has usually practiced the trick before dazzling you with it.) The test of whether you understand a proof is not whether it seems to make sense to you, but whether you can prove the theorem yourself. See whether you can reproduce the professor's proof without looking at it. If you're really feeling enthusiastic, try to prove the same theorem in a different way. Even if you don't succeed, this is great mathematical exercise.

**4.** Most theorems can be proved in several different ways. Your proof of a theorem and my proof of the same theorem need not be the same in order for both to be correct. (In fact, this is one of the most important and enjoyable features of mathematics: that there are many, many ways to arrive at a given mathematical truth.) There are some theorems, though, which are hard to prove unless you remember a particular trick. In this case, you should definitely learn the trick. Memorize it. Practice until you can do it easily, like a card trick. Mathematical tricks are very useful, and you should know as many of them as possible. Your bag of tricks is part of your mathematical toolbox. As Feynman says somewhere, don't despise tricks; make them your own.

5. The following experience is very common among students of mathematics. You stew over a problem for hours or even days, unable to see how to do it. Then, later, you're doing something else and not even thinking about mathematics. Suddenly you see the solution in your mind, and it's obvious.

   This happens to everyone, beginners and professionals alike. The time spent stewing over the problem was not really wasted. It helped you to find the "obvious" solution.

   What is "obvious" depends on what you know. In mathematics, everything you already understand seems easy, and everything you don't understand yet seems impossible.

   A variant of this experience occurs as follows. A professional mathematician is reading a mathematical paper in a scholarly journal. One paragraph contains the sentence: "It is obvious that $x = 0$."

   The mathematician works for three days, and finally proves that $x = 0$.

   "Oh, yes," says the mathematician, without conscious irony. "That *is* obvious."

   You should not use the phrase "it is obvious" in your proofs, even though you may sometimes see this phrase in print.

6. Sometimes you, the student, may feel that everyone else in the class understands the mathematics in this course easily, and that you alone are confused. This impression is almost certainly false. In any case, if you are confused about some detail of a proof that is being presented, you should ask a question immediately. Since mathematics that one doesn't understand seems just like meaningless gibberish, it's always considered polite to ask a question during a mathematical lecture.

If a person at the blackboard keeps writing $x$ where you think $y$ should be, ask about this right away. A person writing on a blackboard needs help from the audience to get the details right. And even if your intended correction is wrong, you need to know that it's wrong in order to understand the rest of the argument.

Often the students who have the least confidence are actually the best students in verbal mathematics. So take heart when you feel that everyone else knows more than you do. It probably isn't true.

7. When you do ask a question in class, you may be disconcerted to find that the professor does not understand what you are asking. This can be frustrating, but it is a perfectly natural occurrence. What we are studying in this class, even more than in verbal mathematics in general, lies right at the threshold of intelligibility. Until your question can be translated into formal mathematical language, its meaning is genuinely unclear. Pronouns such as "it" seem innocuous to the questioner but confuse the hearer. Be patient.

8. A professor who seems irritated by a question is not angry at the questioner. The professor does not understand the question, and is annoyed at not being able to figure it out. Be patient. Professional mathematicians have difficulties akin to those of beginners. Mathematics is not personal, and no one will ever be personally angry with you for asking a mathematical question. The gruffer-seeming mathematicians are not angry; they are just trying to think under pressure, concentrating on the mathematics and not on smiling reassuringly at the student. Gruff professors are much kinder than they seem, and often turn out to be good company.

9. Sometimes the professor makes mistakes, and sometimes mathematical texts contain mistakes. What's great about mathematics is that it makes sense. If you and I make opposing mathematical claims, we can use mathematics to settle which of us (if either) is correct. Don't believe a mathematical statement just because a book or a professor says that it's true and claims to prove it. Is the proof correct? Is the statement true? Can you disprove the statement?

When a professor writing on the blackboard seems to be making a mistake, be sure to ask about it. Whether you're wrong or right, you want to know the truth. One of the most refreshing features of mathematics is that mathematical disagreements are not personal. Any disagreements should be resolved as soon as possible.

10. No methods of thinking, figuring, or computation are beneath the professional mathematician. If we can add better by counting on our fingers, we will do so without shame. If we feel that a picture will help us think about a problem, we draw a picture. (If you would like a professor to draw a picture to illustrate a proof, just ask.)

    Like a magician, a mathematician does plenty of work behind the scenes. When you write a proof for homework, you are performing a magic trick. Your first draft can be messy, with all sorts of false starts and side figuring. Your final draft contains nothing but the finished proof. Looking at your final draft, a reader would think that you had written the proof easily and smoothly, without effort, getting everything right the first time. But you, the magician, know better.

11. Finally, mathematics is one of the great pleasures of the human mind. Even calculating is fun for those who can do it well. The real fun in verbal mathematics comes in seeing something familiar in a new way. When you suddenly understand how an argument works, when you get the joke, that's where the real fun lies.

Here in the U.S.A. at the start of the twenty-first century, most people are completely unacquainted with verbal mathematics. Thus the fun of verbal mathematics has become an unintentional secret. By learning the language of mathematics, you are putting yourself in a position to enjoy some little-known pleasures. Welcome to our secret society.

# THE KID WHO LEARNED ABOUT MATH ON THE STREET

# Chapter 1

# Basic Logic

*When I staggered out from sleep before dawn, I often found her studying calculus at the kitchen table, held in a cloud of Kool smoke like some radiant, unlikely Buddha.*

*"It's a language," she said of the math one morning, tapping her legal pad with the tip of a mechanical pencil. "I've never understood that. It's a language that describes certain stuff very precisely."*

*Mary Carr,* Cherry

*"I know what you're thinking about," said Tweedledum; "but it isn't so, nohow."*

*"Contrariwise," continued Tweedledee, "if it was so, it might be; and if it were so, it would be; but as it isn't, it ain't. That's logic."*

*Lewis Carroll,* Through the Looking Glass

**Introduction.** Logic, or logical reasoning, is a system for drawing conclusions from premises. Premises are the input of a logical argument; the conclusion is the output. Whether the conclusion of an argument is true depends on the truth or falsity of the premises, and on whether or not the conclusion truly follows from the premises.

Since we propose to reason about mathematical objects, we will make use of a formal system of logical operators, called *connectives*. These connectives give us ways to combine statements to obtain other statements. They also give us rules for determining the truth or falsity of the new statements, based on that of the old statements.

Propositional logic concerns true and false statements and logical connectives. The connectives are as follows: *not* ($\neg$), *and* ($\wedge$), *or* ($\vee$), *implies* ($\rightarrow$), and *if and only if* ($\leftrightarrow$). Suppose that $p$ is a statement. If $p$ is true, then $\neg p$ (not $p$) is false. If $p$ is false, then $\neg p$ is true. This tells us all we need to know about the logical connective *not*. Similarly, the connective *and* ($\wedge$) may be described as follows. Let $p$ and $q$ be statements. If $p$ is true and $q$ is true, then $p \wedge q$ ($p$ and $q$) is true. If $p$ is false and $q$ is false, then $p \wedge q$ is false. If $p$ is true and $q$ is false, then $p \wedge q$ is false. If $p$ is false and $q$ is true, then $p \wedge q$ is false.

Since logical connectives are determined by what they do to propositions, we define the connectives by means of truth tables. A truth table tells us what value a connective produces for each possible set of input values. Actually, as we will see in Chapter 4, logical connectives are functions from sets of truth values to sets of truth values. (There are two truth values in our system: *true* and *false.*) But we cannot describe functions between sets without already knowing how to use logical connectives, quantifiers, sets, and so on. Hence, we begin with propositions and connectives defined by means of truth tables. In this first chapter, we will learn the use of logical quantifiers with sets and elements of sets, and the technique of negating quantified statements. By the end of Chapter 1, we will have learned enough formal mathematical language to write and understand mathematical sentences. In the next chapter, we will introduce *definitions* constructed out of words, and will use the definitions and logical reasoning to write paragraphs and prove *theorems.*

We will prove theorems about more and more interesting and complicated mathematical objects as we move through the book and become proficient in the language and lore of mathematics.

But here, at the beginning of Chapter 1, we have not yet even learned to write and interpret formal sentences. So we begin rather inarticulately, with propositions and logical connectives defined by truth tables. By the end of the chapter, we'll be able to start using words to say what we mean in formal mathematical language. Without more ado, let's begin.

**Propositions.** A *proposition*, or statement, is a declarative sentence which is either true or false.

Examples of propositions:

**1.** $3 + 7 = 10$.

**2.** $3 + 7 = 42$.

**3.** $-3 < -4$.

**4.** $5 \geq 7$.

Non-statements:

**1.** $42$.

**2.** $15 - 27$.

**3.** $\sqrt{2} + \sqrt{3}$.

**4.** $x^2$.

**Connectives.** *Connectives*, or logical operators, or truth functions, are defined by means of truth tables. The symbols $p$ and $q$ denote propositions; T stands for "true" and F for "false."

**(a)** *not* ($\neg$)

| $p$ | $\neg p$ |
|---|---|
| T | F |
| F | T |

**(b)** *and* ($\wedge$)

| $p$ | $q$ | $p \wedge q$ |
|---|---|---|
| T | T | T |
| T | F | F |
| F | T | F |
| F | F | F |

**Exercises (1.1)** Construct truth tables for the following propositions.

1. $p \wedge \neg q$

2. $\neg(\neg(\neg p))$

3. $(p \wedge q) \wedge r$

**(c)** *or* ($\vee$)

| $p$ | $q$ | $p \vee q$ |
|---|---|---|
| T | T | T |
| T | F | T |
| F | T | T |
| F | F | F |

This is "the inclusive *or*." Thus $p \vee q$ means: $p$ is true or $q$ is true or both $p$ and $q$ are true.

**Examples (1.1)** The following statements are true.

1. $4 > 5$ or $4 < 5$.
2. $4 \geq 5$ or $4 \leq 5$.
3. $4 > 0$ or $4 < 8$.

**(d)** *if... then* ($\rightarrow$)

| $p$ | $q$ | $p \rightarrow q$ |
|---|---|---|
| T | T | T |
| T | F | F |
| F | T | T |
| F | F | T |

$p \rightarrow q$ means each of the following:

if $p$ then $q$.

$p$ implies $q$.

$p$ only if $q$.

$q$ is a necessary condition for $p$.

$p$ is a sufficient condition for $q$.

**Remark.** In ordinary English, the phrase "if ... then" is used in many different ways. In mathematical English, "if ... then" is used in only one way. Hence the convention that $p \rightarrow q$ is true whenever $p$ is false often strikes people as peculiar.

The mathematical convention concerning "if ... then" corresponds most closely to the use of the phrase when making a promise. Suppose that I say to you, "If it rains tomorrow, I'll invite you to dinner." The only way this statement can be false is that it rains tomorrow and I don't invite you to dinner. In case it doesn't rain tomorrow, my statement to you is true, whether I invite you to dinner or not.

Two propositions are logically equivalent if they have the same truth table.

Example (1.2)

| $p$ | $q$ | $p \to q$ | $\neg(p \to q)$ |
|---|---|---|---|
| T | T | T | F |
| T | F | F | T |
| F | T | T | F |
| F | F | T | F |

| $p$ | $q$ | $\neg q$ | $p \wedge \neg q$ |
|---|---|---|---|
| T | T | F | F |
| T | F | T | T |
| F | T | F | F |
| F | F | T | F |

The tables above show that $\neg(p \to q)$ is logically equivalent to $p \wedge \neg q$.

Exercises (1.2)

1. Show that $p \to q$ is logically equivalent to $\neg p \vee q$.
2. Show that $\neg(p \wedge q)$ is logically equivalent to $\neg p \vee \neg q$.
3. Show that $p \to q$ is logically equivalent to $\neg q \to \neg p$.
4. Classify each of the following propositions as true or false.

   (a) If $5 + 2 = 8$, then $\sqrt{3} = 19$.
   (b) If $5 + 2 = 8$, then $\sqrt{25} = 5$.
   (c) If $5 + 2 = 7$, then $\sqrt{3} = 19$.
   (d) If $5 + 2 = 7$, then $\sqrt{25} = 5$.

**Remark.** Although most people initially find the truth-table definition of $p \to q$ peculiar, this does not result in confusion in practice. It may seem odd that we classify as true such statements as, "If $2 + 3 = 6$ then $2 + 3 = 5$," but, since we really never use such statements in proving theorems, no confusion arises.

**(e)** *if and only if* ($\leftrightarrow$)

| $p$ | $q$ | $p \leftrightarrow q$ |
|---|---|---|
| T | T | T |
| T | F | F |
| F | T | F |
| F | F | T |

$p \leftrightarrow q$ means each of the following:

$p$ if and only if $q$.

$p$ is logically equivalent to $q$.

$(p \rightarrow q) \wedge (q \rightarrow p)$

$p$ and $q$ are either both true or both false.

$p$ is a necessary and sufficient condition for $q$.

$q$ is a necessary and sufficient condition for $p$.

$p$ implies $q$, and $q$ implies $p$.

$(p \rightarrow q) \wedge (\neg p \rightarrow \neg q)$

## Exercises (1.3)

1. Let the symbol # represent "the exclusive *or*." That is, $p \# q$ means: $p$ or $q$ but not both. Write a truth table defining $p \# q$.

2. Use truth tables to show that the following propositions are equivalent:

   **(a)** $p \leftrightarrow q$

   **(b)** $(p \rightarrow q) \wedge (q \rightarrow p)$

   **(c)** $(p \rightarrow q) \wedge (\neg p \rightarrow \neg q)$

**Tautologies and contradictions.** A sentence that is always true is called a *tautology*. For example, $p \vee \neg p$ is a tautology.

| $p$ | $\neg p$ | $p \vee \neg p$ |
|---|---|---|
| T | F | T |
| F | T | T |

A sentence that is always false is called a *contradiction*. For example, $p \wedge \neg p$ is a contradiction.

| $p$ | $\neg p$ | $p \wedge \neg p$ |
|---|---|---|
| T | F | F |
| F | T | F |

In the following set of exercises, the symbol $t$ denotes a tautology and the symbol $c$ denotes a contradiction.

**Exercises (1.4)** Show that the following statements are tautologies. Traditional names for some of these tautologies are given in parentheses. There is no need to memorize these names.

1. $\neg p \vee p$ (Law of Excluded Middle)
2. $p \rightarrow (p \vee q)$ (Addition)
3. $(p \wedge q) \rightarrow p$ (Simplification)
4. $[p \wedge (p \rightarrow q)] \rightarrow q$ (Modus Ponens)
5. $[\neg q \wedge (p \rightarrow q)] \rightarrow \neg p$ (Modus Tollens)
6. $(\neg p \rightarrow c) \rightarrow p$ (Reductio ad Absurdum)

7. $(p \to q) \leftrightarrow [(p \wedge \neg q) \to c]$ (Reductio ad Absurdum)

8. $\neg(p \vee q) \leftrightarrow (\neg p \wedge \neg q)$ (De Morgan's Law)

9. $\neg(p \wedge q) \leftrightarrow (\neg p \vee \neg q)$ (De Morgan's Law)

10. $c \to p$

11. $p \to t$

12. $(p \to q) \leftrightarrow (\neg q \to \neg p)$ (Contrapositive Law)

**Sets.** Loosely speaking, a *set* is a collection of objects. This is not a definition. The notion of a set is basic in mathematics, and the word *set* is not defined. (Since every concept must be defined in terms of concepts whose meaning is already known, some concepts must remain basic and undefined.) Objects which belong to a set are called *elements* of that set.

One way of specifying a set is to list its elements between set brackets.

**Examples (1.3)** The following are examples of sets.

1. $\{1, 3, 7, 9\}$

2. the set of positive even numbers $\{2, 4, 6, 8, \ldots\}$

3. $\{\{1, 2\}, \{1\}\}$

Some well-known sets:

$\mathbb{N}$: the set of natural numbers $\{1, 2, 3, 4, \ldots\}$

$\mathbb{Z}$: the set of integers $\{\ldots -3, -2, -1, 0, 1, 2, 3, \ldots\}$

$\mathbb{Q}$: the set of rational numbers, numbers that can be written as proper or improper fractions

$\mathbb{R}$: the set of real numbers, numbers that can be written as decimals

$\varnothing$: the empty set $\{ \}$

The notation "$a \in A$" means that $a$ is an element of the set $A$. We also say, "$a$ is in $A$," or "$a$ belongs to $A$." To say that $a$ does not belong to $A$, we write "$a \notin A$."

**Examples (1.4)** The following statements are true.

$3 \notin \varnothing$.

$3 \in \mathbb{N}$.

$0 \notin \mathbb{N}$.

$0 \in \mathbb{Z}$.

$1.25 \notin \mathbb{Z}$.

$1.25 \in \mathbb{Q}$.

$\sqrt{2} \notin \mathbb{Q}$.

$\sqrt{2} \in \mathbb{R}$.

$\sqrt{-2} \notin \mathbb{R}$.

**Remarks.** We do not describe the sets $\mathbb{Q}$ and $\mathbb{R}$ more precisely in this chapter because so far we lack the notation to do so. In some books, 0 is regarded as an element of $\mathbb{N}$. In this book, we adhere to the tradition that $0 \notin \mathbb{N}$. We will say more about $\mathbb{Q}$ and $\mathbb{R}$ in Chapter 3.

**Quantifiers.** Quantifiers are important mathematical tools. Using quantifiers, we can make our mathematical language precise. There are two principal quantifiers in mathematics: the universal and the existential.

**Universal quantifier.** The *universal quantifier* has the symbolic form $\forall$. To express the universal quantifier in English, we write "for all," "for every," or "for each."

<u>**Examples**</u> **(1.5)** The following statements are equivalent.

1. $(\forall x \in \mathbb{N})(x+1 \in \mathbb{N})$

2. For every $x$ in $\mathbb{N}$, $x+1$ is in $\mathbb{N}$.

3. For every natural number $x$, the number $x+1$ is also a natural number.

4. Given any natural number $x$, $x+1$ is a natural number.

<u>**Remark.**</u> Of course, there are other ways to state the sentence: "For all $x \in \mathbb{N}$, $x+1 \in \mathbb{N}$." For example, we can say, "If you add 1 to a natural number, you get a natural number." There is nothing wrong with this sentence, but it is not standard "mathematical English." That is, it is not the language of sets and quantifiers. Mathematical diction would sound peculiar in a non-mathematical context. But such language is very useful for expressing mathematical statements.

<u>**Existential quantifier.**</u> The existential quantifier has the symbolic form $\exists$. To express the existential quantifier in words, we say "there exists" or "there is" or "for some" or "there is at least one."

<u>**Examples**</u> **(1.6)** The following statements are equivalent.

1. $(\exists x \in \mathbb{N})(x > 5)$

2. There exists a natural number $x$ such that $x > 5$.

3. There is at least one natural number greater than 5.

4. For some natural number $x$, the number $x$ is greater than 5.

The phrase "there exists [some object]" is often followed by "such that." The phrase "such that" is used in mathematics instead of phrases involving the relative pronouns "which," "that," or "whose."

<u>**Order of quantifiers.**</u> The order of quantifiers in a sentence is important. The following examples illustrate this point.

**Examples (1.7)** Consider these two statements:

1. $(\forall x \in \mathbb{N})(\exists y \in \mathbb{N})(y > x)$

   For each $x \in \mathbb{N}$, there exists $y \in \mathbb{N}$ such that $y > x$.

2. $(\exists y \in \mathbb{N})(\forall x \in \mathbb{N})(y > x)$

   There exists $y \in \mathbb{N}$ such that for each $x \in \mathbb{N}$, $y > x$.

Statement 1 says that given any positive integer, there is a larger positive integer. Statement 1 is true.

Since "$\forall x \in \mathbb{N}$" comes before "$\exists y \in \mathbb{N}$," $y$ depends on $x$. For different values of $x$ there are different values of $y$.

Let $x = 5$. There exists $y \in \mathbb{N}$ such that $y > x$. For example, $6 > 5$.

Let $x = 6$. There exists $y \in \mathbb{N}$ such that $y > 6$. For example, $10 > 6$.

Statement 2 says that there is a positive integer which is larger than every positive integer, including itself.

Statement 2 is false.

Since "$\exists y \in \mathbb{N}$" comes before "$\forall x \in \mathbb{N}$," the value of $y$ does not depend on $x$. The statement says that there is one number $y$ that works for all natural numbers $x$.

**Exercises (1.5)** Write out each statement using words rather than symbols. Then classify the statements either true or false. Explain your answers.

1. $(\exists x \in \mathbb{Z})(\forall y \in \mathbb{Z})(x + y = 0)$
2. $(\forall y \in \mathbb{Z})(\exists x \in \mathbb{Z})(x + y = 0)$
3. $(\forall x \in \mathbb{N})(\exists y \in \mathbb{N})(xy = x)$
4. $(\exists y \in \mathbb{N})(\forall x \in \mathbb{N})(xy = x)$
5. $(\forall x \in \mathbb{N})(\exists y \in \mathbb{N})(x = y - 7)$

6. $(\exists y \in \mathbb{N})(\forall x \in \mathbb{N})(x = y - 7)$

7. $(\forall y \in \mathbb{N})(\exists x \in \mathbb{N})(x = y - 7)$

8. $(\exists x \in \mathbb{N})(\forall y \in \mathbb{N})(x = y - 7)$

**Remarks.** When translating an English sentence into logical symbols, always place a quantifier *before* the statement it governs. English sentences have various ways of expressing quantifiers. For example, consider the sentence: "Any rational number can be expressed as a fraction whose numerator is an integer and whose denominator is a natural number." This sentence can be written symbolically as follows: $(\forall x \in \mathbb{Q})(\exists a \in \mathbb{Z})(\exists b \in \mathbb{N})(x = \frac{a}{b})$. The same sentence can be paraphrased in mathematical English as follows: For each $x \in \mathbb{Q}$, there exist $a \in \mathbb{Z}$ and $b \in \mathbb{N}$ such that $x = \frac{a}{b}$.

The statement $(\forall x \in \mathbb{Q})(\exists a \in \mathbb{Z})(\exists b \in \mathbb{N})(x = \frac{a}{b})$ means the same thing as the statement $(\forall a \in \mathbb{Q})(\exists b \in \mathbb{Z})(\exists x \in \mathbb{N})(a = \frac{b}{x})$. The names of the variables do not matter. Only their roles in the sentence matter.

**Exercises (1.6)** Write each of the following using quantifiers and symbols. In Exercises 9 and 10, the symbol $\varepsilon$ is pronounced "epsilon," with the accent on the first syllable and all vowels short. The symbol $\delta$ is pronounced "delta."

1. For all integers $x$ and $y$, the numbers $xy$ and $yx$ are equal.

2. Given any real number $x$, there exists a natural number $n$ such that $x < n$.

3. Given any real number $x$, there exists a natural number $y$ such that $x + y = 0$.

4. Given any nonnegative real number $x$, there exists a natural number $y$ such that $y^2 = x$.

5. Given any nonzero real number $x$, there exists a natural number $y$ such that $xy = 1$.

6. There exists a smallest natural number.

7. There is no largest integer.

8. Given any two distinct real numbers, some rational number lies strictly between them.

9. Given any positive real number $\varepsilon$, there exists a natural number $k$ such that $\frac{1}{n} < \varepsilon$ whenever $n$ is a natural number greater than $k$.

10. For each real number $\varepsilon$, if $\varepsilon > 0$ then there exists a positive real number $\delta$ such that for each real number $x$, if $|x - 2| < \delta$ then $|x^2 - 4| < \varepsilon$.

**Negating quantified statements.** Let $A$ be a set, and for each $x \in A$ let $p(x)$ be a statement.

Consider statement (a) below.

**(a)** $\neg(\forall x \in A)(p(x))$
It is false that for all $x$ in $A$, $p(x)$ is true.

This statement is equivalent to the following.

**(b)** $(\exists x \in A)(\neg p(x))$
There is at least one $x$ in $A$ for which $p(x)$ is false.

Similarly, statements (c) and (d) are equivalent

**(c)** $\neg(\exists x \in A)(p(x))$
It is false that there is at least one $x$ in $A$ for which $p(x)$ is true.

**(d)** $(\forall x \in A)(\neg p(x))$
For all $x$ in $A$, $p(x)$ is false.

Here is a method for negating quantified sentences. Starting at the beginning of the sentence, change each $\forall$ to $\exists$ and each $\exists$ to $\forall$. Then negate the proposition governed by the quantifier. In verbal sentences, the phrase "such that" is part of the quantifier "there exists." In negating sentences, when we change "there exists" to "for all," the phrase "such that" vanishes with "there exists." When we change "for all" to "there exists," the phrase "such that" appears with "there exists."

**Examples (1.8)**

**1.** Statement: $(\forall x \in \mathbb{N})(x \in \mathbb{Z})$

For each $x \in \mathbb{N}$, $x \in \mathbb{Z}$.

Negation: $(\exists x \in \mathbb{N})(x \notin \mathbb{Z})$

There exists $x \in \mathbb{N}$ such that $x \notin \mathbb{Z}$.

**Remark.** The symbol $\in$ plays two different roles in the sentence $(\forall x \in \mathbb{N})(x \in \mathbb{Z})$. In the expression "$\forall x \in \mathbb{N}$," the symbol $\in$ describes the subject of the sentence. It says that the sentence is about any element $x$ of $\mathbb{N}$. In the phrase "$x \in \mathbb{Z}$," the symbol $\in$ is the verb of the sentence, and tells us that the subject $x$ belongs to the set $\mathbb{Z}$.

**2.** Statement: $(\forall x \in \mathbb{Z})(\exists y \in \mathbb{Z})(x + y = 0)$

For each $x \in \mathbb{Z}$, there exists $y \in \mathbb{Z}$ such that $x + y = 0$.

Negation: $(\exists x \in \mathbb{Z})(\forall y \in \mathbb{Z})(x + y \neq 0)$

There exists $x \in \mathbb{Z}$ such that for all $y \in \mathbb{Z}$, $x + y \neq 0$.

**3.** Statement: $(\forall x \in \mathbb{Q})(\forall y \in \mathbb{Q})((x > y) \to (\exists z \in \mathbb{Q})(x > z > y))$

For each $x \in \mathbb{Q}$, for each $y \in \mathbb{Q}$, if $x > y$ then there exists $z \in \mathbb{Q}$ such that $x > z > y$.

Negation: $(\exists x \in \mathbb{Q})(\exists y \in \mathbb{Q})((x > y) \wedge (\forall z \in \mathbb{Q})((x \leq z) \vee (z \leq y)))$

There exist $x \in \mathbb{Q}$, $y \in \mathbb{Q}$ such that $x > y$ and for all $z \in \mathbb{Q}$, $x \leq z$ or $z \leq y$.

**Remark.** The third example is more complicated than the other two. Hence we offer a step-by-step analysis of the process of negation.

In Example 3, we negate the following statement.

$$(\forall x \in \mathbb{Q})(\forall y \in \mathbb{Q})((x > y) \to (\exists z \in \mathbb{Q})(x > z > y))$$

That is, we produce a statement that is logically equivalent to the following.

$$\neg((\forall x \in \mathbb{Q})(\forall y \in \mathbb{Q})((x > y) \to (\exists z \in \mathbb{Q})(x > z > y)))$$

For our first step, we change the quantifiers at the beginning of the sentence and move the symbol $\neg$ to their right.

$$(\exists x \in \mathbb{Q})(\exists y \in \mathbb{Q})(\neg((x > y) \to (\exists z \in \mathbb{Q})(x > z > y)))$$

Notice that the statement governed by the two initial quantifiers has the form $\neg(p \to q)$. Since $\neg(p \to q)$ is logically equivalent to $p \wedge \neg q$, we obtain the following sentence.

$$(\exists x \in \mathbb{Q})(\exists y \in \mathbb{Q})((x > y) \wedge \neg(\exists z \in \mathbb{Q})(x > z > y))$$

Now we transform the sentence $\neg(\exists z \in \mathbb{Q})(x > z > y)$ by changing the quantifier and moving $\neg$ to the right.

$$(\exists x \in \mathbb{Q})(\exists y \in \mathbb{Q})((x > y) \wedge (\forall z \in \mathbb{Q})(\neg(x > z > y)))$$

Since $x > z > y$ is shorthand for $(x > z) \wedge (z > y)$, we have the following sentence.

$$(\exists x \in \mathbb{Q})(\exists y \in \mathbb{Q})((x > y) \wedge (\forall z \in \mathbb{Q})(\neg((x > z) \wedge (z > y))))$$

Since $\neg(p \wedge q)$ is logically equivalent to $\neg p \vee \neg q$, we transform the sentence as follows.

$$(\exists x \in \mathbb{Q})(\exists y \in \mathbb{Q})((x > y) \wedge (\forall z \in \mathbb{Q})(\neg(x > z) \vee \neg(z > y)))$$

Finally, since $\neg(a > b)$ can be written more simply as $a \leq b$, we get the sentence below.

$$(\exists x \in \mathbb{Q})(\exists y \in \mathbb{Q})((x > y) \wedge (\forall z \in \mathbb{Q})((x \leq z) \vee (z \leq y)))$$

**Remark.** Every step in simplifying a negation moves the symbol $\neg$ further to the right. When all the $\neg$ symbols are as far to the right as possible, we have simplified the negation as much as possible.

**Exercises (1.7)** Classify each statement as either true or false. Also, write two versions of the negation of each statement, one in symbols and one in words. In Exercise 13, recall that $\varepsilon$ is pronounced "epsilon."

**1.** $(\exists x \in \mathbb{R})(x \notin \mathbb{Q})$

**2.** $(\exists x \in \mathbb{N})(\forall y \in \mathbb{N})(y \leq x)$

**3.** $(\forall x \in \mathbb{Z})((x > 0) \rightarrow (x \in \mathbb{N}))$

**4.** $(\forall x \in \mathbb{N})((x > 2) \rightarrow (x > 3))$

**5.** $(\forall x \in \mathbb{N})((\exists k \in \mathbb{N})(x = 2k) \rightarrow (\exists m \in \mathbb{N})(x = 4m))$

**6.** For every natural number $x$, there exists a natural number $y$ such that $x \neq y$ and for every natural number $z$, if $x > z$ then $y > z$.

**7.** For all natural numbers $x$ and $y$, if $x < y$ then there exists a natural number $z$ such that $x < z$ and $z < y$.

**8.** For every integer $x$, either $x$ is a natural number or $-x$ is a natural number.

**9.** For every natural number $x$, there exists a natural number $y$ such that $y^2 = x$.

**10.** For all integers $x$ and $y$, if $x - y$ is a natural number, then $x > y$.

**11.** For each integer $x$, if $|x - 2| > 6$ then $x > 8$ or $x < -4$.

**12.** There exists a natural number $x$ such that for every integer $y$, $x - y$ is a natural number.

**13.** Given any positive real number $\varepsilon$, there exists a natural number $k$ such that $\frac{1}{n} < \varepsilon$ whenever $n$ is a natural number such that $n > k$.

14. For each integer $x$, if $x \leq 0$ then $-x \geq 0$.

15. For each $x \in \mathbb{N}$, there exists $y \in \mathbb{N}$ such that $y = 2x$.

16. For each $x \in \mathbb{N}$, there exists $y \in \mathbb{N}$ such that $x = 2y$.

17. For each $x \in \mathbb{Q}$, if $x \neq 0$ then there exists $y \in \mathbb{Q}$ such that $xy = 1$ and for all $z \in \mathbb{Q}$, if $xz = 1$ then $y = z$.

18. For all real numbers $x$ and $y$, if $x \leq y$ and $x \geq y$ then $x = y$.

19. For each $x \in \mathbb{N}$, if there exists $y \in \mathbb{Q}$ such that $y^2 = x$, then $y \in \mathbb{Z}$.

20. There exists a negative real number $r$ such that $\sqrt{r}$ is real.

21. There exists $x \in \mathbb{N}$ such that for all $y \in \mathbb{N}$, $x \neq 2y$ and $x \neq 2y - 1$.

22. There exists $x \in \mathbb{N}$ such that for all $y \in \mathbb{N}$, if there exists $k \in \mathbb{N}$ such that $yk = x$ then $y = x$ or $y = 1$.

23. There exists $x \in \mathbb{N}$ such that for all $y \in \mathbb{N}$, $x \leq y$.

24. For each real number $x$ such that $x \neq 1$, if $\frac{1}{x-1} > 0$ then $x > 1$.

**Remark.** The preceding sentence may be written, "For each real number $x$, if $x \neq 1$ then if $\frac{1}{x-1} > 0$ then $x > 1$." The phrase *such that* has been used with the universal quantifier to avoid the repetition of *if... then.* (It is also possible to write, "For each real number $x$, if $x \neq 1$ then $\frac{1}{x-1} > 0$ implies $x > 1$.")

25. There exists a real number $x$ such that $x$ is a rational number and $x$ is not a natural number.

26. For every natural number $n$, if $n \neq 1$ then $\frac{1}{n} \notin \mathbb{N}$.

27. There exists $x \in \mathbb{Z}$ such that for all $y \in \mathbb{Z}$, $y^2 \neq x$.

28. For each rational number $x$, the number $x^2$ is also a rational number.

**29.** For each rational number $x$, there exists a rational number $y$ such that $y^2 = x$.

**30.** For all $x \in \mathbb{N}$, if there exists $y \in \mathbb{N}$ such that $y^2 = x$, then for all $z \in \mathbb{N}$, $z^3 \neq x$.

**31.** For all $x \in \mathbb{N}$, there exists $y \in \mathbb{N}$ such that for all $z \in \mathbb{N}$, $z < y$ or $z > x$.

**32.** For all $x \in \mathbb{R}$, if $x \notin \mathbb{Q}$ then for all $k \in \mathbb{Z}$, if $kx \in \mathbb{Q}$ then $k = 0$.

**33.** There exists $x \in \mathbb{R}$ such that for each $y \in \mathbb{Q}$, $y < x$ or $y > x$.

**34.** There exists $x \in \mathbb{R}$ such that $x \neq 0$ and for all $y \in \mathbb{Q}$, if $y > 0$ then $y > x$.

**35.** There exists $x \in \mathbb{R}$ such that $x^2 \notin \mathbb{R}$.

**36.** For all $x, y \in \mathbb{R}$, if $xy = 0$ then $x = 0$ or $y = 0$.

# Chapter 2

# Proving Theorems about Sets

*Schoolmaster: "Suppose $x$ is the number of sheep in the problem."*

*Pupil: "But, Sir, suppose $x$ is not the number of sheep?"*

*(I asked Prof. Wittgenstein was this not a profound philosophical joke, and he said it was.)*

*J. E. Littlewood,* Littlewood's Miscellany

*"She's in that state of mind," said the White Queen, "that she wants to deny* something *— only she doesn't know what to deny!"*

*Lewis Carroll,* Through the Looking Glass

*"Atomics is a very intricate theorem and can be worked out with algebra but you would want to take it by degrees because you might spend the whole night proving a bit of it with rulers and cosines and similar other instruments and then at the wind-up not believe what you had proved at all. If that happened you would have to go back over it till you got to a place where you could believe your own facts and figures as delineated from Hall and Knight's Algebra and then go on again from that particular place till you had the whole thing properly believed and not have bits of it half-believed or a doubt in your head hurting you like when you lose the stud of your shirt in bed."*

*"Very true," I said.*

*Flann O'Brien,* The Third Policeman

**Definition.** Let $A$ and $B$ be sets. $A$ is a *subset* of $B$ if for all $x \in A$, $x \in B$. The symbol $A \subseteq B$ denotes the statement that $A$ is a subset of $B$.

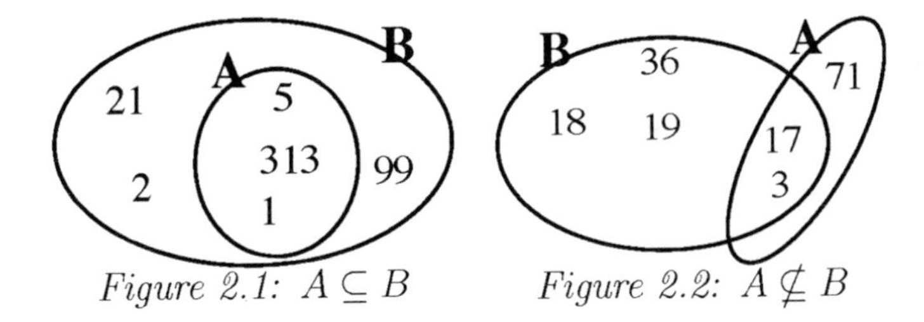

*Figure 2.1:* $A \subseteq B$ *Figure 2.2:* $A \nsubseteq B$

**Remarks.** In mathematical definitions, it is customary to write "if" when we mean "if and only if." Thus the foregoing definition really means that $A \subseteq B$ if and only if, for all $x \in A, x \in B$.

Notice that $A \nsubseteq B$ if and only if there exists $x \in A$ such that $x \notin B$.

**Set builder notation.** Let $A$ be a set, and for all $x \in A$, let $p(x)$ be a proposition about $x$. We can specify a set $S$ as follows: $S = \{x \in A \mid p(x)\}$. The sentence "$S = \{x \in A \mid p(x)\}$" is read aloud as, "$S$ equals the set of all $x$ in $A$ such that $p$ of $x$."

**Examples (2.1)**

1. Let $B = \{1, 2, 3, 4, 5\}$. The set B may be written in set builder notation as $B = \{x \in \mathbb{N} \mid x < 6\}$.

2. Let $E$ be the set of even positive integers. The set $E$ may be written in set builder notation as $E = \{n \in \mathbb{N} \mid (\exists k \in \mathbb{Z})(n = 2k)\}$.

3. The empty set may be written as $\{x \in \mathbb{N} \mid x > 2 \text{ and } x < 2\}$.

**Remark.** In set builder notation, the set $\{x \in A \mid p(x)\}$ is the same as the set $\{y \in A \mid p(y)\}$. Thus, for example, the set $\{n \in \mathbb{N} \mid \text{there exists } k \in \mathbb{N} \text{ such that } n = 3k\}$ is the same as $\{k \in \mathbb{N} \mid \text{there exists } n \in \mathbb{N} \text{ such that } k = 3n\}$. The names of the variables do not matter. The variables in set builder notation are what we call "dummy variables."

**Exercises (2.1)** Describe each set using set builder notation. There is more than one correct answer for each question.

1. $\{1, 7, 9\}$

2. the set of odd positive integers

3. the set of integer multiples of 17

4. $\{4, -4\}$

5. $\{5\}$

6. the set of positive integers greater than 1729

**Remark.** Notice that when defining a new set via set builder notation, the new set is always a subset of a previously defined set. For example, the notation $S = \{x \in \mathbb{N} \mid x \leq 17\}$ introduces the set $S$ as a subset of the familiar set of natural numbers, $\mathbb{N}$. We could define a set $T$ as $\{x \in S \mid x > 5\}$, thus introducing $T$ as a subset of $S$. We do not usually employ such notation as $\{x \mid x > 5\}$, for the following reason.

Although we have not rigorously defined the word "set," not every imaginable aggregation is a set. When Georg Cantor invented set theory in 1895, he mistakenly gave a formal definition of a set as any collection of objects. By considering such entities as “the set of all sets,” Bertrand Russell and others obtained contradictions, thereby showing that there is no such thing as "the set of all sets" and that Cantor’s definition was too general. All this is explained more fully in Afterword A. The upshot is that some care needs to be taken when defining a set. One thing mathematicians have established is that subsets of known sets really are sets. With set builder notation, we define new sets as subsets of known sets. Thus we are assured that the new set we are defining really is a set.

For the same reason, we like to be sure that all the sets in any particular discussion are subsets of some set $X$, which may be called a *universe of discourse* for the discussion. For example, the set of real numbers is a universe of discourse for first-semester calculus, and the set of integers is a universe of discourse for much of number theory. For any particular problem, a universe of discourse is a set $X$ such that all sets involved in the problem are subsets of $X$.

**Definitions.** Let $X$ be a universe of discourse and let $A$ and $B$ be subsets of $X$.

**(i)** The *intersection* of $A$ and $B$, denoted by $A \cap B$, is the set $\{x \in X \mid x \in A \text{ and } x \in B\}$.

**(ii)** The *union* of $A$ and $B$, denoted by $A \cup B$, is the set $\{x \in X \mid x \in A \text{ or } x \in B\}$.

**(iii)** The *relative complement of* $B$ *in* $A$, denoted by $A - B$, is the set $\{x \in X \mid x \in A \text{ and } x \notin B\}$.

**Convention.** Let $X$ be an acknowledged universe of discourse, and let $A$ be a subset of $X$. The relative complement of $A$ in $X$ is often simply called the *complement of* $A$ and denoted by $A^C$.

**Example (2.2)** Let $X = \{1, 3, 5, 7, 9\}$. Let $A = \{1, 3, 5\}$. Let $B = \{5, 9\}$. Then $A \cap B = \{5\}$, $A \cup B = \{1, 3, 5, 9\}$, $A - B = \{1, 3\}$, $B - A = \{9\}$, $A^C = \{7, 9\}$, and $B^C = \{1, 3, 7\}$.

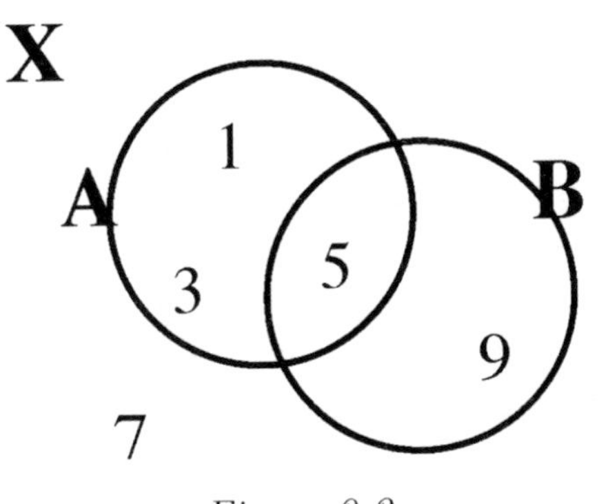


*Figure 2.3*

**Definition.** Let $A$ and $B$ be sets. The sets $A$ and $B$ are *equal* if $A \subseteq B$ and $B \subseteq A$.

We are now ready to begin proving theorems about sets. Our first proof is an example of a proof by contradiction, or indirect proof.

**Theorem** **(2.1)** For every set $A$, $\varnothing \subseteq A$.

**Proof.** By way of contradiction, suppose that there exists a set $A$ such that $\varnothing \nsubseteq A$. Since $\varnothing \nsubseteq A$, there exists $x \in \varnothing$ such that $x \notin A$. But $\varnothing$ has no elements, so $x \notin \varnothing$. Hence $x \in \varnothing$ and $x \notin \varnothing$. $\rightarrow\leftarrow$

Our hypothesis has led to a contradiction and is therefore false. Thus for every set $A$, $\varnothing \subseteq A$. Q.E.D.

**Remarks.** We begin our final draft by writing down the word *Theorem*, followed by a period or colon. Then we write the statement of the theorem. On the next line, we write *Proof*, followed by a period or colon. Now we are ready to write the proof.

To write a proof by contradiction, we begin by assuming that the proposition we wish to prove is false. Then we argue logically until we reach a contradiction, a statement that must be false. (For example, $x \in \varnothing$ and $x \notin \varnothing$.) The symbol $\rightarrow\leftarrow$ means that we have arrived at a contradiction.

The proposition $(\neg p \rightarrow c) \rightarrow p$ is a tautology. To write an indirect proof, we assume $\neg p$ and show that $\neg p \rightarrow c$, where $c$ is a contradiction. This proves the proposition $p$.

The letters Q.E.D. stand for *quod erat demonstrandum.* This is Latin, and means "which was to be proved," or, more colloquially, "which is what we set out to prove in the first place." In Latin, *quod erat demonstrandum* may stand alone as a sentence meaning "This is what was to be proved." It means that the proof is finished.

The beginning of a proof by contradiction is important. We write, "By way of contradiction, suppose..." and then write the *negation* of the proposition we are trying to prove. We are trying to prove: For every set $A$, $\varnothing \subseteq A$. The negation of this proposition is: There exists a set $A$

such that $\varnothing \not\subseteq A$. Thus we begin by writing: "By way of contradiction, suppose that there exists a set $A$ such that $\varnothing \not\subseteq A$."

We have introduced a set $A$ such that $\varnothing \not\subseteq A$. We now use the definition of "subset." Recall that $(\varnothing \not\subseteq A) \leftrightarrow (\exists x \in \varnothing)(x \notin A)$. We write, "Since $\varnothing \not\subseteq A$, there exists $x \in \varnothing$ such that $x \notin A$."

We have now introduced an $x$ which is an element of the empty set. But this cannot be. The empty set has no elements. We have reached a contradiction.

The proposition $p$ that we wish to prove states that for every set $A$, the empty set $\varnothing$ is a subset of $A$. We began: "By way of contradiction, suppose $\neg p$." Thus $\neg p$ is our hypothesis. We have shown that $\neg p$ implies a contradiction. This proves that $p$ is true.

Proofs by contradiction are like jokes. We assume something untrue and argue logically until we reach a conclusion we recognize as ridiculous. Sometimes the conclusion is quite startling.

We now present another proof by contradiction. This time, the statement of the theorem is of the form $p \to q$.

**Theorem (2.2)** For all sets $A$ and $B$, if $A \subseteq A \cap B$ then $A \cup B \subseteq B$.

**Proof.** By way of contradiction, suppose that there exist sets $A$ and $B$ such that $A \subseteq A \cap B$ and $A \cup B \not\subseteq B$.

Since $A \cup B \not\subseteq B$, there exists $x \in A \cup B$ such that $x \notin B$. Since $x \in A \cup B$, $x \in A$ or $x \in B$.

Case 1. Suppose that $x \in A$. Since $x \in A$ and $A \subseteq A \cap B$, $x \in A \cap B$. Since $x \in A \cap B$, $x \in A$ and $x \in B$. Thus $x \in B$. But by the choice of $x$, $x \notin B$. $\to\leftarrow$

Case 2. Suppose that $x \in B$. By the choice of $x$, $x \notin B$. Thus $x \in B$ and $x \notin B$. $\to\leftarrow$

In each case, our hypothesis has led to a contradiction and is therefore false. Thus, for all sets $A$ and $B$, if $A \subseteq A \cap B$ then $A \cup B \subseteq B$. Q.E.D.

**Remarks.** We wish to prove that for all sets $A$ and $B$, if $A \subseteq A \cap B$ then $A \cup B \subseteq B$. The negation of this proposition is: There exist sets $A$ and $B$ such that $A \subseteq A \cap B$ and $A \cup B \not\subseteq B$. So we begin by writing, "By way of contradiction, suppose that there exist sets $A$ and $B$ such that $A \subseteq A \cap B$ and $A \cup B \not\subseteq B$."

We proceed on the assumption that $A \subseteq A \cap B$ and $A \cup B \not\subseteq B$. We recall that $(A \cup B \not\subseteq B) \leftrightarrow (\exists x \in A \cup B)(x \notin B)$. We write, "Since $A \cup B \not\subseteq B$, there exists $x \in A \cup B$ such that $x \notin B$."

In Case 1, we use the definition of "subset" to conclude that since $x \in A$ and $A \subseteq A \cap B$, it follows that $x \in A \cap B$.

Notice that we must finish both Case 1 and Case 2 before we can conclude that our hypothesis (that there exist sets $A$ and $B$ such that $A \subseteq A \cap B$ and $A \cup B \not\subseteq B$) is contradictory. One case alone is not enough to allow us to draw this conclusion.

**Exercises (2.2)** For each theorem, write the opening lines of a proof by contradiction.

1. Theorem. For all sets $A$ and $B$, if $A \subseteq B$ then $A - B = \varnothing$.

2. Theorem. For all sets $A$, $B$, and $C$, if $A \subseteq C$ and $B \subseteq C$ then $A \cup B \subseteq C$.

3. Theorem. For all sets $A$, $B$, and $C$, $A \cap (B \cup C) \subseteq (A \cap B) \cup (A \cap C)$.

4. Theorem. For all sets $A$, $B$, and $C$, if $A \cap C \subseteq B \cap C$ and $A \cup C \subseteq B \cup C$ then $A \subseteq B$.

**Remark.** Proving theorems involves thought and reflection. It is an activity that should not be done frantically or in a hurry. When you prove a theorem, always write down the statement of the theorem first. You are not wasting time when you write out mathematics by hand; you are giving your mind a chance to learn and understand on its own terms. Most of mathematical thinking is done by the unconscious mind. We don't know how this works. We do know that writing mathematical statements out by hand helps the mind to learn.

**Exercises (2.3)**

**1 – 4.** Write a proof by contradiction of each theorem from Exercises 2.1. Begin by writing down the statement of the theorem. Use the opening lines you wrote for Exercises 2.2.

For each theorem, write a proof by contradiction.

**5.** Theorem. Let $A$, $B$, and $X$ be sets. If $A \subseteq B$, then $X - B \subseteq X - A$.

**6.** Theorem. Let $A$, $B$, and $X$ be sets. If $A \cup B \subseteq X$ and $X - B \subseteq X - A$, then $A \subseteq B$.

**7.** Theorem. For all sets $A$, $B$, and $C$, $(A \cap B) \cup (A \cap C) \subseteq A \cap (B \cup C)$.

**8.** Theorem. For all sets $A$, $B$, and $C$, if $A \subseteq B$ and $A \subseteq C$, then $A \subseteq B \cap C$.

Now we present an example of a direct proof.

**Theorem (2.3)** For all sets $A$, $B$, and $C$, $(A \cap B) \cup (A \cap C) \subseteq A \cap (B \cup C)$.

**Proof.** Let $A$, $B$, and $C$ be sets. Let $x \in (A \cap B) \cup (A \cap C)$. Since $x \in (A \cap B) \cup (A \cap C)$, it follows that $x \in A \cap B$ or $x \in A \cap C$.

Case 1. Suppose $x \in A \cap B$. Since $x \in A \cap B$, $x \in A$ and $x \in B$. Since $x \in B$, $x \in B \cup C$. Since $x \in A$ and $x \in B \cup C$, $x \in A \cap (B \cup C)$.

Case 2. Suppose $x \in A \cap C$. Since $x \in A \cap C$, $x \in A$ and $x \in C$. Since $x \in C$, $x \in B \cup C$. Since $x \in A$ and $x \in B \cup C$, $x \in A \cap (B \cup C)$.

Thus, for all $x \in (A \cap B) \cup (A \cap C)$, $x \in A \cap (B \cup C)$.

Therefore, $(A \cap B) \cup (A \cap C) \subseteq A \cap (B \cup C)$. Q.E.D.

**Remarks.** Let's examine the structure of this direct proof in detail. We wish to prove that for any three sets $A$, $B$, and $C$, $(A \cap B) \cup (A \cap C) \subseteq A \cap (B \cup C)$.

Sometimes beginners are inclined to write in generalizations, beginning each sentence with, "For all sets $A$, $B$, and $C, \ldots$" This is not a good way to write a proof. Instead, we introduce the characters in our drama by saying, "Let $A$, $B$, and $C$ be sets." Whenever we want to discuss a new mathematical object, we must introduce it with a sentence beginning: "Let..."

Thus, the first sentence of our proof is, "Let $A$, $B$, and $C$ be sets." This tells everyone what we are talking about. We may, if we wish, follow up this introduction with a statement of our intentions: "We will show that $(A \cap B) \cup (A \cap C) \subseteq A \cap (B \cup C)$." This statement of intention is not strictly necessary, but it can be helpful, for it reminds everyone of what we are trying to prove.

We have now introduced three sets, $A$, $B$, and $C$, as characters in the drama of our proof. But we don't know anything about these sets, except that they are sets. Sets $A$, $B$, and $C$ might all be the same set. One or more of them might be empty. We don't know.

Now that we have introduced the sets $A$, $B$, and $C$, we must show that $(A \cap B) \cup (A \cap C) \subseteq A \cap (B \cup C)$. That is, according to the definition of "subset," we must show that for all $x \in (A \cap B) \cup (A \cap C)$, $x \in A \cap (B \cup C)$. Hence we begin, "Let $x \in (A \cap B) \cup (A \cap C)$." If we wish, we may append the optional statement of intention, "We will show that $x \in A \cap (B \cup C)$."

We have written: "Let $x \in (A \cap B) \cup (A \cap C)$." According to our definition of union, $(A \cap B) \cup (A \cap C) = \{y \in X \mid y \in A \cap B \text{ or } y \in A \cap C\}$ where $X$ is a universe of discourse. That is, since $x \in (A \cap B) \cup (A \cap C)$, by definition $x \in A \cap B$ or $x \in A \cap C$.

We have written "$x \in A \cap B$ or $x \in A \cap C$." The unwary novice often proceeds, "Therefore, $x \in A$ and $x \in B$ or $x \in A$ and $x \in C$. So $x \in A$

or $x \in A$ and $x \in A$ or $x \in C$ and $x \in B$ or $x \in A$ and $x \in B$ or $x \in C$."
This is ambiguous and hard to follow.

Instead, we divide our argument into cases when we reach a proposition of the form "$p$ or $q$." We write: "Case 1: Suppose that $p$." After finishing Case 1, we write "Case 2: Suppose that $q$."

Since we wish to show that $x \in A \cap (B \cup C)$, the conclusion of each case is "$x \in A \cap (B \cup C)$." In each case, to arrive at the desired conclusion we use the definitions of union and intersection step by step and give a reason for each step. For example, when we write, "Since $x \in A \cap B$, $x \in A$ and $x \in B$," we are asserting, "$x \in A$ and $x \in B$," and the reason given is "$x \in A \cap B$." Here the meaning of the word "since" is "because."

After the cases are complete, we state what we have shown so far: "Thus, for all $x \in (A \cap B) \cup (A \cap C)$, $x \in A \cap (B \cup C)$." We conclude by writing: "Therefore, $(A \cap B) \cup (A \cap C) \subseteq A \cap (B \cup C)$. Q.E.D." Omitting the conclusion of a proof is like telling a joke and leaving out the punch line.

These remarks are aimed at describing the structure of our proof of Theorem 2.3. Our finished proof does not include the myriad details discussed in these remarks. However, we can, and should, write down anything we like on our scratch paper, our rough draft. We count on our fingers, draw pictures, work out examples, and investigate ideas that seem too silly even to mention. But we don't include all this in our final copy. For the final copy, we just write the finished proof, as if we had pulled it by magic out of thin air.

Theorem 2.3 may be stated in either of two ways:

**(a)** For all sets $A$, $B$, and $C$, $(A \cap B) \cup (A \cap C) \subseteq A \cap (B \cup C)$.

**(b)** Let $A$, $B$, and $C$ be sets. Then $(A \cap B) \cup (A \cap C) \subseteq A \cap (B \cup C)$.

Since (a) and (b) mean the same thing, they are proved in the same way.

Similarly, if $X$ is a set and for each $x \in X$, $p(x)$ is a proposition about $x$, then the statement, "For all $x \in X$, $p(x)$ is true," means the same thing as the two statements: "Let $x \in X$. Then $p(x)$ is true."

Let $p$ be the proposition $x \in A$, let $q$ be the proposition $x \in B$, and let $r$ be the proposition $x \in C$. Proving Theorem 2.3 amounts to proving $((p \wedge q) \vee (p \wedge r)) \rightarrow (p \wedge (q \vee r))$. We can prove this using truth tables. Why not do it that way?

Truth tables are used to define the logical connectives $\neg$, $\wedge$, $\vee$, $\rightarrow$, and $\leftrightarrow$. We never use them in writing mathematical proofs. Proofs by truth table are neither interesting nor easy to follow, unless what is being proved is very simple indeed. Moreover, truth-table proofs do not tell us very much. They can tell us that a certain proposition is true, but not why it is true.

When we write a proof that for all sets $A$, $B$, and $C$, $(A \cap B) \cup (A \cap C) \subseteq A \cap (B \cup C)$, we are thinking in terms of sets. We may draw pictures to convince ourselves:

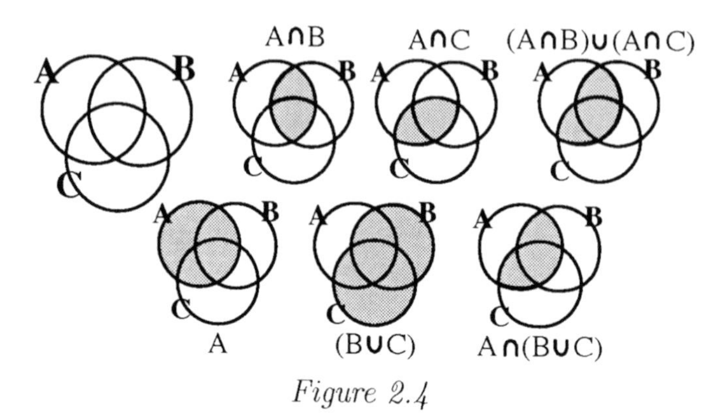


*Figure 2.4*

We may think of examples. For instance, let $A = \{2, 4, 6\}$, let $B = \{1, 3, 6\}$, and let $C = \{3, 5\}$, and check to make sure that the theorem holds in this case.

When writing a proof about sets, you should certainly feel free to draw pictures and to work out examples. However, the pictures and examples do not constitute a proof of a statement about *all* sets $A$, $B$, and $C$. The pictures and examples are part of the magician's fiddling behind the scenes. They help you, the mathematician, to think, to understand the statement, and to confirm that the statement is probably true before proving it. They may show you how to prove the theorem. But they are not part of the finished proof.

Why not accept a picture or an example as proof of a theorem? Pictures and examples can be misleading. Consider the example given above: $A = \{2, 4, 6\}$, $B = \{1, 3, 6\}$, and $C = \{3, 5\}$. We can represent this situation in a diagram thus:

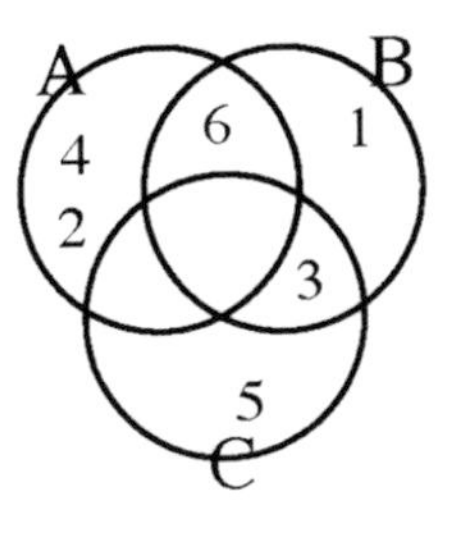


*Figure 2.5*

For these particular sets, $A \cap B \cap C = \varnothing$. This does not mean that for all sets $A$, $B$, and $C$, $A \cap B \cap C = \varnothing$. It does mean that there exist sets $A$, $B$, and $C$ such that $A \cap B \cap C = \varnothing$.

The following two sentences are not equivalent:

(a) For all sets $A$, $B$, and $C$, $(A \cap B) \cup (A \cap C) \subseteq A \cap (B \cup C)$.

(b) There exist sets $A$, $B$, and $C$ such that $(A \cap B) \cup (A \cap C) \subseteq A \cap (B \cup C)$.

A picture or an example proves only statement (b). A proof of Theorem 2.3 proves the much stronger statement (a).

There is certainly a sense in which a picture can sometimes be regarded as a rigorous proof. But at this point in our study of mathematics, we are learning to use words. When you present a proof at the blackboard, by all means include any sketches that will help the reader to follow the proof. But the proof itself will consist of words.

**Exercises (2.4)** Write a direct proof of each theorem.

1. Theorem: Let $A$, $B$, and $X$ be sets such that $A \subseteq B$. Then $X - B \subseteq X - A$.

2. Theorem: Let $A$, $B$, and $X$ be sets. Then $(X - A) \cup (X - B) \subseteq X - (A \cap B)$.

3. Theorem: Let $A$, $B$, and $C$ be sets such that $A \subseteq B$ and $B \subseteq C$. Then $A \subseteq C$.

4. Theorem: For all sets $A$, $B$, and $C$, $A \cup (B \cap C) \subseteq (A \cup B) \cap (A \cup C)$.

Now let's examine a direct proof of a theorem of the form $p \rightarrow q$.

**Theorem (2.4)** For all sets $A$ and $B$, if $A \cup B \subseteq B$ then $A \subseteq A \cap B$.

**Proof.** Let $A$ and $B$ be sets such that $A \cup B \subseteq B$. Let $x \in A$. Since $x \in A$, it follows that $x \in A \cup B$. Since $x \in A \cup B$ and $A \cup B \subseteq B$, $x \in B$. Since $x \in A$ and $x \in B$, $x \in A \cap B$.

Thus for all $x \in A$, $x \in A \cap B$. That is, $A \subseteq A \cap B$.

Therefore, for all sets $A$ and $B$, if $A \cup B \subseteq B$ then $A \subseteq A \cap B$. Q.E.D.

**Remarks.** This theorem has the form of an if-then statement: for all sets $A$ and $B$, $p \rightarrow q$. We recall that whenever $p$ is false the proposition $p \rightarrow q$ is automatically true. Thus, we need not consider the case where $p$ is false. We begin: "Let $A$ and $B$ be sets such that $p$ is true." If we wish, we may append the statement of intention: "We will show that $q$ is true."

We wish to show that $A \subseteq A \cap B$. That is, we wish to show that for all $x \in A$, $x \in A \cap B$.

Hence, our next statement is: "Let $x \in A$." If we wish, we may append the optional statement of intention: "We will show that $x \in A \cap B$."

We then proceed step by step using the definitions of union, subset, and intersection to arrive at the desired conclusion. Each step follows logically from previous statements, and the steps are strung together like beads on a string, with the conclusion as the last bead.

In general, the middle part of the proof is the most difficult (and also the most entertaining) because it may very well not be obvious which steps to take or in what order. When in doubt how to proceed, use scratch paper to try out an idea, and if that idea doesn't work, try something else. It's not unusual for several pages of scratch paper to be used to produce a short one-paragraph proof. (See Dr. Spencer's Mantra for ideas for what to do when you get stuck. Bear in mind that getting stuck is a natural part of the process of proving theorems. We get stuck. Then we get blissfully unstuck. Then we get stuck on something else. And so on.)

Finally, don't leave out the punchline. Be sure to finish the proof by writing the conclusion.

**Exercises (2.5)** For each theorem, write the opening lines of a direct proof.

1. Theorem. For all sets $A$ and $B$, if $A = A \cap B$ then $A \subseteq B$.

2. Theorem. For all sets $A$ and $B$, if $A \cup B = B$ then $A \subseteq B$.

3. Theorem. For all sets $A$, $B$, $C$, and $D$, if $A \subseteq C$ and $B \subseteq D$ then $A \cup B \subseteq C \cup D$.

4. Theorem. For all sets $A$, $B$, and $C$, if $A \cap C \subseteq B \cap C$ and $A \cup C \subseteq B \cup C$ then $A \subseteq B$.

**Exercises (2.6)** For each theorem in Exercises 2.5, write a direct proof. Be sure to write out the statement of the theorem before proving it. Use the opening lines you composed for Exercises 2.5.

## Two-way proofs: set equality and logical equivalence.

1. To prove that two sets $A$ and $B$ are equal, prove both that $A \subseteq B$ and that $B \subseteq A$.

2. To prove $p \leftrightarrow q$, prove both $p \to q$ and $q \to p$. (Or prove both $p \to q$ and $\neg p \to \neg q$.)

The next theorem is a proof of the equality of two sets.

**Theorem (2.5)** For all sets $A$ and $B$, $(A \cap B)^C = A^C \cup B^C$.

**Proof.** Let $A$ and $B$ be sets.

First we will show that $(A \cap B)^C \subseteq A^C \cup B^C$.

Let $x \in (A \cap B)^C$. Then $x \notin A \cap B$. Since $x \notin A \cap B$, $x \notin A$ or $x \notin B$.

Case 1. Suppose $x \notin A$. Then $x \in A^C$. Since $x \in A^C$, $x \in A^C \cup B^C$.

Case 2. Suppose $x \notin B$. Then $x \in B^C$. Since $x \in B^C$, $x \in A^C \cup B^C$.

Thus, for all $x \in (A \cap B)^C$, $x \in A^C \cup B^C$.

Hence $(A \cap B)^C \subseteq A^C \cup B^C$.

It remains to show that $A^C \cup B^C \subseteq (A \cap B)^C$.

Let $x \in A^C \cup B^C$. Then $x \in A^C$ or $x \in B^C$.

Case 1. Suppose $x \in A^C$. Since $x \in A^C$, $x \notin A$. Since $x \notin A$, $x \notin A \cap B$. Since $x \notin A \cap B$, $x \in (A \cap B)^C$.

Case 2. Suppose $x \in B^C$. Since $x \in B^C$, $x \notin B$. Since $x \notin B$, $x \notin A \cap B$. Since $x \notin A \cap B$, $x \in (A \cap B)^C$.

Thus, for all $x \in A^C \cup B^C$, $x \in (A \cap B)^C$.

Hence $A^C \cup B^C \subseteq (A \cap B)^C$.

Since $(A \cap B)^C \subseteq A^C \cup B^C$ and $A^C \cup B^C \subseteq (A \cap B)^C$, $(A \cap B)^C = A^C \cup B^C$. Q.E.D.

We now present a proof of the logical equivalence of two propositions.

**Theorem (2.6)** For all sets $A$ and $B$, the set $A \cap B$ is empty if and only if $A \subseteq B^C$.

**Proof.** Let $A$ and $B$ be sets. First we will show that if $A \cap B = \varnothing$ then $A \subseteq B^C$.

Suppose $A \cap B = \varnothing$. Let $x \in A$. Since $A \cap B = \varnothing$, it follows that $x \notin A \cap B$. Since $x \notin A \cap B$ and $x \in A$, $x \notin B$. Since $x \notin B$, $x \in B^C$.

Thus for all $x \in A$, $x \in B^C$. That is, $A \subseteq B^C$.

Therefore, if $A \cap B = \varnothing$ then $A \subseteq B^C$.

It remains to show that if $A \subseteq B^C$ then $A \cap B = \varnothing$.

By way of contradiction, suppose that $A \subseteq B^C$ and $A \cap B \neq \varnothing$. Since $A \cap B \neq \varnothing$, there exists $x \in A \cap B$. Since $x \in A \cap B$, $x \in A$ and $x \in B$. Since $x \in B$, $x \notin B^C$. Thus $x \in A$ and $x \notin B^C$. But since $x \in A$ and $A \subseteq B^C$, it follows that $x \in B^C$. Thus $x \in B^C$ and $x \notin B^C$. $\rightarrow\leftarrow$

Our hypothesis is contradictory and therefore false. Therefore, if $A \subseteq B^C$ then $A \cap B = \varnothing$.

Since $A \cap B = \varnothing$ implies $A \subseteq B^C$ and since $A \subseteq B^C$ implies $A \cap B = \varnothing$, $A \cap B = \varnothing$ if and only if $A \subseteq B^C$. Q.E.D.

**Exercises (2.7)** Prove the following theorems.

1. Theorem. Let $A$ and $B$ be sets. Then $A - B = \varnothing$ if and only if $A \subseteq B$.

2. Theorem. For all sets $A$ and $B$, $(A - B) \cup (B - A) = (A \cup B) - (A \cap B)$.

3. Theorem. Let $A$, $B$, and $X$ be sets. If $A \cup B \subseteq X$ then $A - B = (X - B) - (X - A)$.

**False Propositions and counterexamples.** We cannot prove just any old proposition about sets. Some propositions are true and some are false. If we can prove a proposition, then it is true.

How do we prove that a false proposition is false? Most theorems are written as universal statements: For all sets $A$, $B$, and $C$, the statement $p$ is true. To disprove such a universal statement, we must exhibit sets $A$, $B$, and $C$ such that the statement $p$ is false. We call this *finding a counterexample.*

To disprove the false universal statement $(\forall x \in A)(p(x))$, first write the negation of the statement. This gives $(\exists x \in A)(\neg p(x))$. To provide a counterexample, we must exhibit an element $x$ of $A$ such that $\neg p(x)$.

We now offer an illustration of a disproof of a false universal statement.

**False Proposition.** For all sets $A$, $B$, and $C$, if $A \subseteq B \cup C$ then $A \subseteq B$ or $A \subseteq C$.

**Counterexample.** We will show that there exist sets $A$, $B$, and $C$ such that $A \subseteq B \cup C$ and $A \not\subseteq B$ and $A \not\subseteq C$. Let $A = \{1, 2, 3\}$, let $B = \{1, 2\}$, and let $C = \{3\}$. Then $B \cup C = \{1, 2, 3\}$. Since $A = B \cup C$, $A \subseteq B \cup C$. But $A \not\subseteq B$, because $3 \in A$ and $3 \notin B$. Also, $A \not\subseteq C$, because $1 \in A$ and $1 \notin C$. Thus $A$, $B$, and $C$ are sets such that $A \subseteq B \cup C$ but $A \not\subseteq B$ and $A \not\subseteq C$. Hence the proposition above is false.

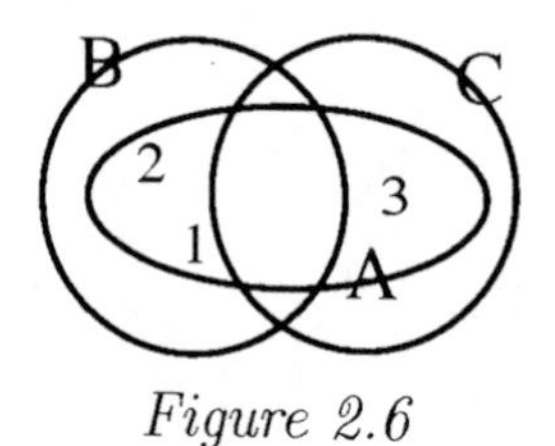


*Figure 2.6*

**Remarks.** The proposition we wish to disprove states that for all sets $A$, $B$, and $C$, if $A \subseteq B \cup C$ then $A \subseteq B$ or $A \subseteq C$. We negate this proposition, and obtain: There exist sets $A$, $B$ and $C$ such that $A \subseteq B \cup C$ and $A \not\subseteq B$ and $A \not\subseteq C$. To provide a counterexample, we need to find such sets $A$, $B$, and $C$.

One counterexample is enough. We just need to find three sets $A$, $B$, and $C$ that do the trick. A simple counterexample is best, because it is easy for the reader to understand. In this case, we chose the sets $A = \{1,2,3\}$, $B = \{1,2\}$, and $C = \{3\}$. These are not the only possible choices.

Now that we have chosen sets $A$, $B$, and $C$ to disprove the proposition, we must prove that these three sets constitute a counterexample. That is, we must prove that $A \subseteq B \cup C$ and $A \not\subseteq B$ and $A \not\subseteq C$.

**Exercises (2.8)** For each false proposition write a disproof by counterexample.

1. False Proposition. For all sets $A$ and $B$, $A \cup B = (A - B) \cup (B - A)$.
2. False Proposition. For all sets $A$ and $B$, $A \cap B \neq A - B$.
3. False Proposition. For all sets $A$ and $B$, if $A \cup B = A$ then $B = \varnothing$.
4. False Proposition. For all sets $A$, $B$, and $C$, $A \cap (B \cup C) = (A \cap B) \cup C$.

Some of the statements below are theorems and some are false propositions. Prove each theorem and disprove each false proposition.

5. For all sets $A$ and $B$, if $A \not\subseteq B$ then $A \cap B = \varnothing$.
6. For all sets $A$, $B$, and $C$, if $C \cap (A \cap B) = \varnothing$ then $C \cap A = \varnothing$ or $C \cap B = \varnothing$.
7. For all sets $A$, $B$, and $X$, $X - (A \cup B) = (X - A) \cap (X - B)$.
8. For all sets $A$, $B$, and $C$, if $C \subseteq A$ and $C \subseteq B$ then $C \subseteq A \cap B$.
9. For all sets $A$, $B$, and $C$, if $(A \cap B) \cup C \subseteq A \cap (B \cup C)$ then $C \subseteq A$.
10. For all sets $A$, $B$, and $X$, if $A \subseteq B$ then $X - A \subseteq X - B$.
11. For all sets $A$, $B$, and $C$, if $A \cup C \subseteq B \cup C$ then $A \subseteq B$.
12. For all sets $A$, $B$, and $C$, if $A \subseteq B$ then $A \cup C \subseteq B \cup C$.

**13.** For all sets $A$, $B$, and $C$, if $A \subseteq B$ then $A - C \subseteq B - C$.

**14.** For all sets $A$, $B$, and $C$, $A - (B - C) = (A - B) - C$.

**15.** For all sets $A$, $B$, and $C$, $(A - B) - C = (A - C) - (B - C)$.

<u>Exercises</u> **(2.9)** The following theorems are well-known and frequently used in all areas of mathematics. Prove the theorems. (We have already proved some of them.)

1. <u>Theorem</u>. For each set $A$, $\varnothing \subseteq A$.

2. <u>Theorem</u>. For each set $A$, $A \cap A = A$.

3. <u>Theorem</u>. For each set $A$, $A \cup A = A$.

4. <u>Theorem</u>. For each set $A$, $A = A$.

5. <u>Theorem</u>. For each set $A$, $A \cup \varnothing = A$.

6. <u>Theorem</u>. For each set $A$, $A \cap \varnothing = \varnothing$.

7. <u>Theorem</u>. For all sets $A$ and $B$, $A \cap B = B \cap A$.

8. <u>Theorem</u>. For all sets $A$ and $B$, $A \cup B = B \cup A$.

9. <u>Theorem</u>. For all sets $A$, $B$, and $C$, $(A \cup B) \cup C = A \cup (B \cup C)$.

10. <u>Theorem</u>. For all sets $A$, $B$, and $C$, $(A \cap B) \cap C = A \cap (B \cap C)$.

11. <u>Theorem</u>. For all sets $A$ and $B$, $A \subseteq B$ if and only if $A \cap B = A$.

12. <u>Theorem</u>. For all sets $A$ and $B$, $B \subseteq A$ if and only if $B - A = \varnothing$.

13. <u>Theorem</u>. For all sets $A$ and $B$, $A \subseteq B$ if and only if $B^C \subseteq A^C$.

14. <u>Theorem</u>. For all sets $A$ and $B$, $(A \cup B)^C = A^C \cap B^C$. [DeMorgan's Law]

15. <u>Theorem</u>. For all sets $A$ and $B$, $(A \cap B)^C = A^C \cup B^C$. [DeMorgan's Law]

**16.** Theorem. For all sets $A$ and $B$, $A - B = B^C - A^C$.

**17.** Theorem. For all sets $A$, $B$, and $C$, if $A \subseteq B$ and $B \subseteq C$ then $A \subseteq C$. [Transitive Law]

**18.** Theorem. For all sets $A$, $B$, and $C$, if $A \subseteq B$ and $A \subseteq C$ then $A \subseteq B \cap C$.

**19.** Theorem. For all sets $A$, $B$, and $C$, if $A \subseteq C$ and $B \subseteq C$ then $A \cup B \subseteq C$.

**20.** Theorem. For all sets $A$, $B$, and $C$, $A \cap (B \cup C) = (A \cap B) \cup (A \cap C)$. [Distributive Law]

**21.** Theorem. For all sets $A$, $B$, and $C$, $A \cup (B \cap C) = (A \cup B) \cap (A \cup C)$. [Distributive Law]

# Chapter 3

# Cartesian Products and Relations

*On visiting Cambridge the Master (Montague Butler) asked him at a dinner whether he was related to "our dear Ernest Harrison." Adopting a certain philosophical view of "relations" (repudiated by Russell) he replied: — No.*

*J.E. Littlewood,* Littlewood's Miscellany

*— define the number in terms of the equivalence class of ordered pairs of the equivalence class of ordered pairs . . .*

*— Anybody know what he's talking about?*

*William Gaddis,* JR

*Of such was Neary's love for Miss Dwyer, who loved a Flight Lieutenant Elliman, who loved a Miss Farren of Ringsakiddy, who loved a Father Fitt of Ballinclashet, who in all sincerity was bound to acknowledge a certain vocation for a Mrs. West of Passage, who loved Neary.*

*Samuel Beckett,* Murphy

Before proceeding further, we will say a few words about the sets $\mathbb{Q}$ and $\mathbb{R}$.

**Definition.** The set of *rational numbers* $\mathbb{Q}$ is the set $\{\frac{p}{q} \mid p \in \mathbb{N}$ and $q \in \mathbb{Z}\}$.

**Remark.** The set $\mathbb{Q}$ is the set of all numbers $x$ which can be expressed as fractions. Thus $3 \in \mathbb{Q}$, because $3 = \frac{3}{1}$ (or $3 = \frac{24}{8}$). Whether a number belongs to $\mathbb{Q}$ or not does not depend on how it is written.

**Examples (3.1)** The numbers 486, $-7.2$, $\frac{3}{4}$, 0 and $-46\frac{2}{3}$ all belong to $\mathbb{Q}$. The numbers $\sqrt{2}$, $-\sqrt{3}$, $\pi$, $e$, and $56.1011011101111011111\ldots$ (where the length of each successive string of 1's exceeds the length of the preceding string by 1) do not belong to $\mathbb{Q}$. These are real numbers but not rational numbers. They are called *irrational numbers.*

**The real numbers.** The set $\mathbb{R}$ of *real numbers* is the set of all numbers on the number line. The set $\mathbb{R}$ contains both the rational and the irrational numbers. All decimal numbers are real numbers.

**Remarks.** We do not define the real numbers at this point because we still lack the proper notation for even a naive definition. We will become more precise as we develop our notation.

A student once said that he had been told the following three things: (a) $\pi = 3.14$; (b) $\pi = \frac{22}{7}$; and (c) $\pi$ is irrational. How, he wanted to know, could all three things be true?

The number $\pi$ is irrational. Therefore, $\pi$ cannot be expressed precisely as a fraction or a decimal number of finite length. The beginning of the decimal expansion of $\pi$ is $3.1415926535897932\ldots$ The expansion goes on forever. The numbers 3.14 and $\frac{22}{7}$ are rational numbers used to approximate the number $\pi$, but $\pi \neq 3.14$ and $\pi \neq \frac{22}{7}$.

**Ordered pairs.** Let $X$ be a set, and let $a, b \in X$. Loosely speaking, the ordered pair $(a, b)$ consists of the elements $a$ and $b$, in that order. (This is not a definition.) Thus, for all $a, b, c, d \in X$, $(a, b) = (c, d)$ if and only if $a = c$ and $b = d$.

**Remark.** Notice that the ordered pair $(a, b)$ is not the same object as the set $\{a, b\}$. If $a \neq b$, then $\{a, b\} = \{b, a\}$ but $(a, b) \neq (b, a)$.

**Definition.** Let $A$ and $B$ be sets. The *cross product* of $A$ and $B$, denoted by $A \times B$, is the set $\{(a, b) \mid a \in A \text{ and } b \in B\}$. The cross product of $A$ and $B$ may also be called the *Cartesian product*, or simply the *product*, of $A$ and $B$.

**Example (3.2)** Let $A = \{2, 1\}$ and let $B = \{2, 8, 3\}$. Then $A \times B = \{(2, 2), (2, 8), (2, 3), (1, 2), (1, 8), (1, 3)\}$.

| $A \setminus B$ | 2 | 8 | 3 |
|---|---|---|---|
| 2 | (2, 2) | (2, 8) | (2, 3) |
| 1 | (1, 2) | (1, 8) | (1, 3) |

$A \times B$

**Notation.** Let $A$ be a set. The set $A \times A$ may be denoted by $A^2$.

**Example (3.3)** The set $\mathbb{R} \times \mathbb{R}$ is usually denoted by $\mathbb{R}^2$. The symbol $\mathbb{R}^2$ is usually pronounced "are two," not "are squared."

**Exercises (3.1)**

1. Let $A = \{27, 47, 83\}$ and let $B = \{67, 89\}$. Find $A \times B$.

   In Exercises 2 through 7, prove the stated theorem.

2. Theorem. For each set $A$, $A \times \varnothing = \varnothing$ and $\varnothing \times A = \varnothing$.

3. Theorem. For all sets $A$ and $B$, $A \times B = \varnothing$ if and only if $A = \varnothing$ or $B = \varnothing$.

4. Theorem. For all sets $A$, $B$, and $C$, $(A \cup B) \times C = (A \times C) \cup (B \times C)$.

5. Theorem. For all sets $A$, $B$, $C$, and $D$, $(A \cap B) \times (C \cap D) = (A \times C) \cap (B \times D)$.

6. Theorem. For all sets $A$, $B$, and $C$, $(A - B) \times C = (A \times C) - (B \times C)$.

7. Theorem. For all sets $A$, $B$, $C$, and $D$, if $A \subseteq B$ and $C \subseteq D$, then $A \times C \subseteq B \times D$.

In Exercises 8 and 9, provide a counterexample to disprove the false proposition.

8. False Proposition. For all sets $A$, $B$, $C$, and $D$, if $A \times C \subseteq B \times D$, then $A \subseteq B$ and $C \subseteq D$.

9. False Proposition. For all nonempty sets $A$, $B$, and $C$, if $C \subseteq A \times B$ then there exist sets $E \subseteq A$ and $F \subseteq B$ such that $C = E \times F$.

10. In formal, axiomatic set theory (as opposed to the naive or intuitive set theory of this book), all objects of discussion are defined as sets. One formal system defines the ordered pair $(a, b)$ to be the set $\{\{a\}, \{a, b\}\}$. Using this definition, prove the following theorem.

    Theorem. Let $X$ be a set, and let $a, b, c, d \in X$. Then $(a, b) = (c, d)$ if and only if $a = c$ and $b = d$.

**Definition.** Let $x \in \mathbb{N}$ and let $y \in \mathbb{Z}$. Then $x$ *divides* $y$ if there exists $k \in \mathbb{Z}$ such that $y = kx$. The symbol $x \mid y$ denotes the statement that $x$ divides $y$.

**Remark.** The following statements are equivalent: (a) $x$ divides $y$; (b) $y$ is divisible by $x$; (c) $x$ is a divisor of $y$; (d) $x$ is a factor of $y$; (e) $y$ is a multiple of $x$.

**Examples (3.4)**

1. $2 \mid 6$, because $2 \times 3 = 6$.

2. $10 \mid -30$, because $(10)(-3) = -30$.

3. $5 \mid 0$, because $5 \times 0 = 0$.

4. $2 \nmid 7$, because for all $x \in \mathbb{Z}$, $2x \neq 7$.

5. $3 \nmid 5$, because for all $x \in \mathbb{Z}$, $3x \neq 5$.

**Definition.** Let $n \in \mathbb{N}$ and let $x, y \in \mathbb{Z}$. Then $x$ *is congruent to* $y$ *modulo* $n$ if there exists $k \in \mathbb{Z}$ such that $x - y = kn$. The notation $x \equiv y \mod n$ means that $x$ is congruent to $y$ modulo $n$.

**Examples (3.5)**

1. $3 \equiv 7 \mod 4$, because $3 - 7 = (-1)(4)$.

2. $155 \equiv 1 \mod 11$, because $155 - 1 = (14)(11)$.

3. $-8 \equiv 7 \mod 5$, because $-8 - 7 = (-3)(5)$.

4. $6 \equiv -1 \mod 7$, because $6 - (-1) = (1)(7)$.

**Definitions.** Let $A$ and $B$ be sets. A *relation from* $A$ *to* $B$ is a subset of $A \times B$. Let $A$ be a set. A *relation on* $A$ is a relation from $A$ to $A$.

**Notation.** Let $A$ and $B$ be sets, and let $\boldsymbol{r}$ be a relation from $A$ to $B$. Let $a \in A$ and $b \in B$. Instead of $(a, b) \in \boldsymbol{r}$, we usually write $a\,\boldsymbol{r}\,b$. Instead of "$(a, b)$ is in $\boldsymbol{r}$," we say "$a$ is related by $\boldsymbol{r}$ to $b$."

**Examples (3.6)** The following are examples of relations.

1. $\boldsymbol{r} = \{(x, y) \in \mathbb{R}^2 \mid x = y\}$.

2. $\boldsymbol{r} = \{(x, y) \in \mathbb{R}^2 \mid x < y\}$.

3. $\boldsymbol{r} = \{(a, b) \in \mathbb{R}^2 \mid b = a^2\}$.

4. $\boldsymbol{r} = \{(a, b) \in \mathbb{N}^2 \mid b = 2a + 1\}$.

5. $r$ on $\mathbb{Z}$ such that $a\, r\, b$ if and only if $a + b = 10$.

6. $r$ on $\mathbb{Z}$ such that $a\, r\, b$ if and only if $a \neq b$.

7. $r$ on $\mathbb{R}^2$ such that $(x, y)\, r\, (u, v)$ if and only if $x + y \leq u + v$.

8. $r$ on $\mathbb{N}$ such that $a\, r\, b$ if and only if $a \mid b$.

9. $r$ on $\mathbb{Z}$ such that $a\, r\, b$ if and only if $a \equiv b \bmod 17$.

**Remarks.** Although a relation $r$ is defined as a set of ordered pairs, we do not usually think of it this way. We seldom use relations as nouns (that is, as names of sets). More often, relations are used as verbs or predicates.

For instance, consider examples 1, 2, and 8 above. Instead of "$(x, y) \in r$," we say, in Example 1, "$x$ equals $y$," in Example 2, "$x$ is less than $y$," and in Example 8, "$x$ divides $y$."

Instead of thinking of a relation as a set of ordered pairs, we think of a relation as a system of connections between pairs of objects, usually expressed in sentences. Thus we write "$1 \leq 2$," not "$(1, 2) \in \leq$."

**Example (3.7)** Let $A$ be the set $\{1, 3, 5, 7\}$ and let $B$ be the set $\{2, 5, 8, 11\}$. Let $r$ be the relation $\{(x, y) \in A \times B \mid x > y\}$.

We seldom think of $r$ as a set of ordered pairs, like this.

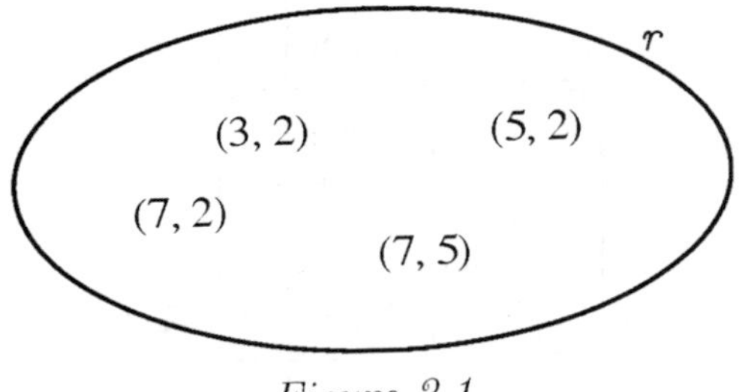


*Figure 3.1*

We think of $r$ as a system of connections between elements of $A$ and elements of $B$, like this.

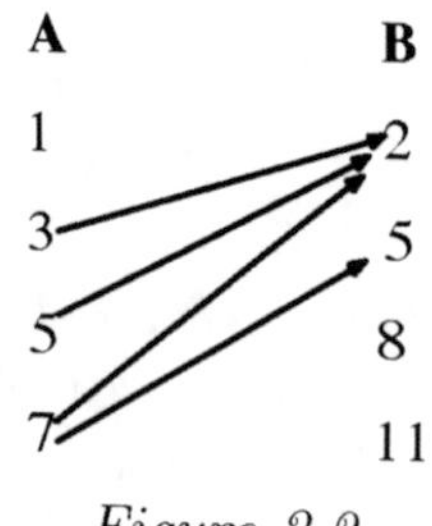


*Figure 3.2*

We may represent $r$ as a sequence of sentences.

$$3 > 2. \quad 5 > 2. \quad 7 > 2. \quad 7 > 5.$$

*Figure 3.3*

We may even represent $r$ as a matrix, like this.

| $A \setminus B$ | 2 | 5 | 8 | 11 |
|---|---|---|---|---|
| 1 | 0 | 0 | 0 | 0 |
| 3 | 1 | 0 | 0 | 0 |
| 5 | 1 | 0 | 0 | 0 |
| 7 | 1 | 1 | 0 | 0 |

*Figure 3.4*

(Here a 1 in a row $a$, column $b$, means that $a\,\boldsymbol{r}\,b$; a 0 means that $(a, b) \notin \boldsymbol{r}$.)

In writing proofs, we usually express relations in sentences, as in Figure 3.3 above. When we are thinking about relations, however, we are free to use any model that helps us think. Sometimes it is helpful to use the formal definition of a relation from $A$ to $B$ as a subset of $A \times B$. Usually, however, we find it most convenient to express relations in sentences when we write proofs.

In mathematics, we often make use of several ways of looking at a particular mathematical object. It is very useful (as well as entertaining) to be able to represent the same object in several different ways. The ability to see one object in many different ways is an important skill in mathematics, and well worth cultivating.

**Definitions.** Let $A$ and $B$ be sets, and let $\boldsymbol{r}$ be a relation from $A$ to $B$.

**(i)** The *domain* of $\boldsymbol{r}$, denoted by Dom $\boldsymbol{r}$, is the set $\{a \in A \mid$ there exists $b \in B$ such that $a\,\boldsymbol{r}\,b\}$.

**(ii)** The *image* of $\boldsymbol{r}$, denoted by Im $\boldsymbol{r}$, is the set $\{b \in B \mid$ there exists $a \in A$ such that $a\,\boldsymbol{r}\,b\}$. The image of $\boldsymbol{r}$ is also called the *range* of $\boldsymbol{r}$.

**Example (3.8)** Let $\boldsymbol{r} = \{(a, b) \in \mathbb{N}^2 \mid a = b^3\}$. Then $\operatorname{Dom} \boldsymbol{r} = \{1, 8, 27, 64, 125 \ldots\} = \{n \in \mathbb{N} \mid$ there exists $m \in \mathbb{N}$ such that $m^3 = n\}$, whereas $\operatorname{Im} \boldsymbol{r} = \mathbb{N}$.

**Exercises (3.2)**

1. For each relation in Examples 3.6, list three elements of $\boldsymbol{r}$.

2. For each relation $\boldsymbol{r}$ in Examples 3.6, find Dom $\boldsymbol{r}$ and Im $\boldsymbol{r}$.

**Definitions.** Let $A$ be a set, and let $\boldsymbol{r}$ be a relation on $A$.

**(i)** The relation $\boldsymbol{r}$ is *reflexive* if for all $x \in A$, $x\,\boldsymbol{r}\,x$.

**(ii)** The relation $\boldsymbol{r}$ is *symmetric* if for all $x, y \in A$, $x\,\boldsymbol{r}\,y$ implies $y\,\boldsymbol{r}\,x$.

**(iii)** The relation $\boldsymbol{r}$ is *transitive* provided that for all $x, y, z \in A$, if $x\,\boldsymbol{r}\,y$ and $y\,\boldsymbol{r}\,z$ then $x\,\boldsymbol{r}\,z$.

**(iv)** The relation $\boldsymbol{r}$ is an *equivalence relation* if $\boldsymbol{r}$ is reflexive, symmetric, and transitive.

**Remark.** In the definition of *transitive* above, "provided that" means "if." We have used this phrase to avoid repeating "if."

**Theorem (3.1)** Let $\boldsymbol{r}$ be the relation on $\mathbb{Z}$ such that $a\,\boldsymbol{r}\,b$ if and only if $a^2 = b^2$. Then $\boldsymbol{r}$ is an equivalence relation on $\mathbb{Z}$.

**Proof.** We will show that $\boldsymbol{r}$ is reflexive, symmetric, and transitive.

Let $a \in \mathbb{Z}$. Since $a^2 = a^2$, $a\,\boldsymbol{r}\,a$. Hence, for all $a \in \mathbb{Z}$, $a\,\boldsymbol{r}\,a$. Thus $\boldsymbol{r}$ is reflexive.

Let $a$, $b \in \mathbb{Z}$. Suppose that $a\,\boldsymbol{r}\,b$. Then $a^2 = b^2$. Thus $b^2 = a^2$, and so $b\,\boldsymbol{r}\,a$. Hence, for all $a$, $b \in \mathbb{Z}$, if $a\,\boldsymbol{r}\,b$ then $b\,\boldsymbol{r}\,a$. Thus $\boldsymbol{r}$ is symmetric.

Let $a$, $b$, $c \in \mathbb{Z}$. Suppose that $a\,\boldsymbol{r}\,b$ and $b\,\boldsymbol{r}\,c$. Then $a^2 = b^2$ and $b^2 = c^2$. Hence $a^2 = c^2$. Thus $a\,\boldsymbol{r}\,c$. Hence, for all $a$, $b$, $c \in \mathbb{Z}$, if $a\,\boldsymbol{r}\,b$ and $b\,\boldsymbol{r}\,c$ then $a\,\boldsymbol{r}\,c$. Thus, $\boldsymbol{r}$ is transitive.

Since $\boldsymbol{r}$ is reflexive, symmetric, and transitive, $\boldsymbol{r}$ is an equivalence relation on $\mathbb{Z}$. Q.E.D.

**Theorem (3.2)** The relation $<$ on the set of real numbers is not reflexive, is not symmetric, and is transitive.

**Proof.** Since $1 \not< 1$, there exists $x \in \mathbb{R}$ (namely, $x = 1$) such that $x \not< x$. Hence, $<$ is not reflexive.

Since $1 < 2$ but $2 \not< 1$, there exist $x$, $y \in \mathbb{R}$ (namely, $x = 1$ and $y = 2$) such that $x < y$ but $y \not< x$. Hence, $<$ is not symmetric.

Let $x$, $y$, $z \in \mathbb{R}$. Suppose that $x < y$ and $y < z$. Since $x < y$ and $y < z$, $x < z$. Hence, for all $x$, $y$, $z \in \mathbb{R}$, if $x < y$ and $y < z$, then $x < z$. Thus, $<$ is transitive.

Therefore, the relation $<$ on $\mathbb{R}$ is not reflexive, is not symmetric, and is transitive. Q.E.D.

**Exercises (3.3)** In problems 1 through 13, determine whether or not the given relation is: (a) reflexive; (b) symmetric; (c) transitive. Prove that your answers are correct.

1. $r = \{(a, b) \in \mathbb{R}^2 \mid b = a^2\}$.

2. $r = \{(a, b) \in \mathbb{N}^2 \mid b = 2a + 1\}$.

3. $r$ on $\mathbb{Z}$ such that $a\, r\, b$ if and only if $a + b = 10$.

4. $r$ on $\mathbb{Z}$ such that $a\, r\, b$ if and only if $a \neq b$.

5. $r$ on $\mathbb{R}^2$ such that $(x, y)\, r\, (u, v)$ if and only if $x + y \leq u + v$.

6. $r = \{(a, b) \in \mathbb{N}^2 \mid a \mid b\}$.

7. Congruence modulo 7, defined on $\mathbb{Z}$.

8. $r$ on $\mathbb{R}$ such that $x\, r\, y$ if and only if $|x - 4| = |y - 4|$.

9. $r$ on $\mathbb{R}^2$ such that $(u, v)\, r\, (x, y)$ if and only if $u^2 + v^2 = x^2 + y^2$.

10. $r$ on $\mathbb{R}^2$ such that $(u, v)\, r\, (x, y)$ if and only if $y - x^2 = v - u^2$

11. $r$ on $\mathbb{R}^2$ such that $(u, v)\, r\, (x, y)$ if and only if $v - 2u = y - 2x$.

12. $r$ on $\mathbb{R}^2$ such that $(u, v)\, r\, (x, y)$ if and only if $uv = xy$.

13. $r$ on $\mathbb{Z} \times \mathbb{N}$ such that $(u, v)\, r\, (x, y)$ if and only if $uy = vx$.

In problems 14 through 18, prove the given theorem.

14. Theorem. For each $n \in \mathbb{N}$, congruence modulo $n$ is an equivalence relation on $\mathbb{Z}$.

15. Theorem. Let $n \in \mathbb{N}$, and let $a, b, c \in \mathbb{Z}$. If $a \equiv b \bmod n$, then $a + c \equiv b + c \bmod n$.

16. Theorem. Let $n \in \mathbb{N}$, and let $a, b, c, d \in \mathbb{Z}$. If $a \equiv b \bmod n$ and $c \equiv d \bmod n$, then $a + c \equiv b + d \bmod n$.

17. Theorem. Let $n \in \mathbb{N}$, and let $a, b, c \in \mathbb{Z}$. If $a \equiv b \bmod n$, then $ac \equiv bc \bmod n$.

18. Theorem. Let $n \in \mathbb{N}$, and let $a, b, c, d \in \mathbb{Z}$. If $a \equiv b \bmod n$ and $c \equiv d \bmod n$, then $ac \equiv bd \bmod n$.

**Definition.** Let $A$ be a set, let $x \in A$, and let $\sim$ be an equivalence relation on $A$. The *equivalence class* of $x$ under $\sim$ is the set $[x] = \{y \in A \mid x \sim y\}$. The symbol $[x]$ denotes the equivalence class of $x$ under an equivalence relation $\sim$ .

**Definition.** Let $A$ be a set and let $\sim$ be an equivalence relation on $A$. Then $A$ *modulo* $\sim$ is the set $\{[x] \mid x \in A\}$. In other words, $A$ modulo $\sim$ is the set of equivalence classes under the relation $\sim$. The symbol $A/\sim$ denotes the set $A$ modulo $\sim$ .

**Example (3.9)** Let $\sim$ be the equivalence relation on $\mathbb{N}$ of congruence modulo 3. Then $\mathbb{N}/\sim = \{\{1, 4, 7, 10, \ldots\}, \{2, 5, 8, 11, \ldots\}, \{3, 6, 9, 12, \ldots\}\} = \{[1], [2], [3]\}$.

**Example (3.10)** Let $\sim$ be the equivalence relation on $\mathbb{R}^2$ such that $(u, v) \sim (x, y)$ if and only if $u^2 + v^2 = x^2 + y^2$. Then the equivalence class $[(3, 0)]$ under $\sim$ is the set $\{(x, y) \in \mathbb{R}^2 \mid x^2 + y^2 = 9\}$. If we think of the set $\mathbb{R}^2$ as the familiar Cartesian plane with the usual $x$ and $y$ axes, then the set $[(3, 0)]$ is the set of points on the graph of the circle with center $(0, 0)$ and radius 3.

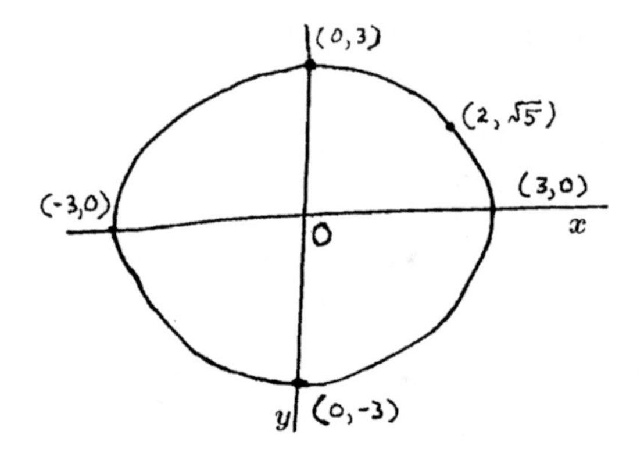


*Figure 3.5:* $[(3, 0)] = \{(x, y) \in \mathbb{R}^2 \mid x^2 + y^2 = 9\}$

For each $u, v \in \mathbb{R}$, $[(u, v)] = \{(x, y) \in \mathbb{R}^2 \mid x^2 + y^2 = u^2 + v^2\}$. Thus $[(0, 0)] = \{(0, 0)\}$, and for all $u, v \in \mathbb{R}$, if $u \neq 0$ or $v \neq 0$ then $[(u, v)]$ is the set of points on the circle with center $(0, 0)$ and radius $\sqrt{u^2 + v^2}$.

The set $\mathbb{R}^2/\!\sim\; = \{[(0, a)] \mid a \in \mathbb{R}$ and $a \geq 0\}$. For each positive real number $a$, let $C_a$ be the circle in $\mathbb{R}^2$ with center $(0, 0)$ and radius $a$. (That is, $C_a = \{(x, y) \in \mathbb{R}^2 \mid x^2 + y^2 = a^2\}$.) Then for each positive real number $a$, $[(0, a)] = C_a$. Thus the elements of $\mathbb{R}^2/\!\sim$ are the circles in the plane centered at the origin, including the "degenerate circle" of radius 0.

When we say that the set $\mathbb{R}^2/\!\sim\; = \{[(0, a)] \mid a \in \mathbb{R}$ and $a \geq 0\}$, we are specifying the set $\mathbb{R}^2/\!\sim$ of equivalence classes under $\sim$ by naming one element of each equivalence class. For each $a \in \mathbb{R}$, if $a \geq 0$, then the equivalence class $[(0, a)]$ is the same as the class $[(a, 0)] = [(-a, 0)]$ $= [(\frac{a}{\sqrt{2}}, \frac{a}{\sqrt{2}})] = [(\frac{a}{2}, \frac{-a\sqrt{3}}{2})]$. Thus, instead of "$\mathbb{R}^2/\!\sim\; = \{[(0, a)] \mid a \in \mathbb{R}$ and $a \geq 0\}$," we could equally well say, "$\mathbb{R}^2/\!\sim\; = \{[(\frac{a}{2}, \frac{-a\sqrt{3}}{2})] \mid a \in \mathbb{R}$ and $a \geq 0\}$." We have chosen the formulation we find most convenient.

**Exercises (3.4)** For each equivalence relation given below, find and list all the equivalence classes. In Exercises 8, 9, and 10, also sketch the graph of three of the equivalence classes. (The product $\mathbb{R}^2$ is the Cartesian plane. Graph three equivalence classes in the plane $\mathbb{R}^2$ using the usual horizontal $x$-axis and vertical $y$-axis.)

**1.** $r$ on $A = \{1, 2, 3\}$ such that $r = \{(1, 1), (2, 2), (3, 3), (1, 2), (2, 1)\}$.

**2.** $r$ on $\mathbb{Z}$ such that $a\, r\, b$ if and only if $a^2 = b^2$.

**3.** Congruence modulo 2 on $\mathbb{Z}$.

**4.** Congruence modulo 3 on $\mathbb{Z}$.

**5.** Congruence modulo 10 on $\mathbb{Z}$.

**6.** Congruence modulo $n$ on $\mathbb{Z}$.

**7.** The relation $r$ on $\mathbb{R}$ such that $x\, r\, y$ if and only if $|x-4| = |y-4|$.

8. The relation $r$ on $\mathbb{R}^2$ such that $(u, v)\, r\, (x, y)$ if and only if $y - x^2 = v - u^2$.

9. The relation $r$ on $\mathbb{R}^2$ such that $(u, v)\, r\, (x, y)$ if and only if $v - 2u = y - 2x$.

10. The relation $r$ on $\mathbb{R}^2$ such that $(u, v)\, r\, (x, y)$ if and only if $uv = xy$.

**Definition.** Let $X$ be a set, and let $a, b \in X$. The elements $a$ and $b$ are *distinct* if $a \neq b$.

**Definition.** Let $A$ and $B$ be sets. The sets $A$ and $B$ are *disjoint* if $A \cap B = \varnothing$.

**Notation: sets used to index other sets.** There is a joke about a mathematician telling a bedtime story to a child. "Once upon a time," says the mathematician, "there existed three bears: $B_1$, $B_2$, and $B_3$."

The mathematician is using the set $\{1, 2, 3\}$ to index the set of bears. The subscripts 1, 2, and 3 are called *indices* (plural of index).

The joke could be carried further; the mathematician might say, "Once upon a time there existed three distinct bears: $B_1$, $B_2$, and $B_3$." With the word *distinct*, the mathematician indicates that the three bears are all different from one another, that $B_1$, $B_2$, and $B_3$ are not just three different names for the same bear, or two names for one bear and one name for another.

At the risk of ceasing to be funny, let's expand the joke about the bedtime story. "Once upon a time," says the mathematician, "there existed three distinct bears: $B_1$, $B_2$, and $B_3$, who lived in a house $H$ in the forest $F$. One day the three bears went for a walk. When they came home, they found that all their beds had been slept in.

"Bear $B_1$ said: There exists a trespasser $T_1$ such that $T_1$ has slept in my bed, $b_1$.

"Bear $B_2$ said: There exists a trespasser $T_2$ such that $T_2$ has slept in $b_2$.

"Bear $B_3$ said: There exists a trespasser $T_3$ such that $T_3$ has slept in $b_3$.

"The three bears thought long and hard about the collection of trespassers $\mathscr{T} = \{T_i\}_{i=1}^{3}$."

Eventually, the bears will discover that $T_1 = T_2 = T_3$. For each $i \in \{1, 2, 3\}$, the trespasser $T_i$ is a little girl named Goldilocks.

Considered as a set, $\mathscr{T}$ has only one element. (The symbol $\mathscr{T}$ may be pronounced "script $T$.")

Considered as an indexed set, or sequence, $\mathscr{T}$ has three terms.

There are several ways to denote an indexed set. Let $A$ be a set, and let $G$ be a collection indexed by $A$. We may write $G = \{x_a \mid a \in A\}$ or $G = \{x_a\}_{a \in A}$. In the special case where the indexing set $A$ is the set $\mathbb{N}$ of natural numbers, we usually write $G = \{x_n\}_{n=1}^{\infty}$. This notation is equivalent to: $G = \{x_n\}_{n \in \mathbb{N}}$. In this special case the indexed set $G$ is sometimes called a *sequence*, or *infinite sequence*. In the case where for some $n \in \mathbb{N}$ the indexing set $A$ is the finite set $\{k \in \mathbb{N} \mid k \leq n\}$, we often write $G = \{x_k\}_{k=1}^{n}$ or $G = \{x_1, x_2, \ldots, x_n\}$ instead of $G = \{x_a\}_{a \in A}$. Such a finite indexed set $G$ is called a *finite sequence*.

Sometimes the elements of an indexed set are sets themselves. In this case, the indexed set is often denoted by a cursive capital letter such as $\mathscr{G}$, $\mathscr{Q}$, $\mathscr{F}$, etc.

The words "collection" and "family" are synonyms of "set." Usually sets whose elements are themselves sets are called "families of sets" or "collections of sets," rather than "sets of sets."

**Examples (3.11)**

1. The sequence $\{x_n\}_{n=1}^{\infty}$ where for each $n \in \mathbb{N}$, $x_n = (-1)^n$.

2. The finite sequence $\{a_k\}_{k=1}^{5}$ where for each $1 \leq k \leq 5$, $a_k = \frac{k^2+k}{2}$.

3. The collection $\mathscr{G} = \{G_n\}_{n \in \mathbb{N}}$ where $G_n = \{n, \ n+1\}$.

4. The finite collection $\mathscr{F} = \{F_i\}_{i=1}^{10}$ where $F_i = \{n \in \mathbb{N} \mid n \leq \ i\}$.

**Remarks.** The phrase "for each $1 \leq k \leq 5$" in Example 2 is standard shorthand for: "for each $k \in \mathbb{N}$ such that $1 \leq k \leq 5$."

Mathematicians often use Greek letters to denote indexing sets and indecies. Thus, this seems a natural point at which to introduce the Greek alphabet. The use of Greek letters is widespread in mathematics, and is by no means restricted to naming indexing sets and their elements. In general, mathematicians do not know Greek, but they do know the names of the Greek letters. You should feel free to make use of Greek letters when you run out of Latin (English) letters. The best Greek letters to use, of course, are those which look different from Latin letters. (The letter omicron, for example, is not useful in mathematics.)

The Greek alphabet is as follows:

| | | | |
|---|---|---|---|
| A $\alpha$ alpha | H $\eta$ eta | N $\nu$ nu | T $\tau$ tau |
| B $\beta$ beta | $\Theta$ $\theta$ theta | $\Xi$ $\xi$ xi | $\Upsilon$ $\upsilon$ upsilon |
| $\Gamma$ $\gamma$ gamma | I $\iota$ iota | O o omicron | $\Phi$ $\phi$ phi |
| $\Delta$ $\delta$ delta | K $\kappa$ kappa | $\Pi$ $\pi$ pi | X $\chi$ chi |
| E $\varepsilon$ epsilon | $\Lambda$ $\lambda$ lambda | P $\rho$ rho | $\Psi$ $\psi$ psi |
| Z $\zeta$ zeta | M $\mu$ mu | $\Sigma$ $\sigma$ sigma | $\Omega$ $\omega$ omega |

**Definitions.** Let $X$ be a universe of discourse. Let $\Gamma$ be a set, and for each $\gamma \in \Gamma$ let $A_\gamma$ be a set. Let $\mathscr{A} = \{A_\gamma\}_{\gamma \in \Gamma}$.

(i) The *union* of $\mathscr{A}$, or the union of the elements of $\mathscr{A}$, is the set $\{x \in X \mid$ there exists $\gamma \in \Gamma$ such that $x \in A_\gamma\}$. The union of the elements of $\mathscr{A}$ is denoted by $\bigcup_{\gamma \in \Gamma} A_\gamma$ or by $\bigcup \mathscr{A}$.

(ii) The *intersection* of $\mathscr{A}$, or the intersection of the elements of $\mathscr{A}$, is the set $\{x \in X \mid$ for all $\gamma \in \Gamma$, $x \in A_\gamma\}$. The intersection of the elements of $\mathscr{A}$ is denoted by $\bigcap_{\gamma \in \Gamma} A_\gamma$ or by $\bigcap \mathscr{A}$.

**Example (3.12)** Let $\Gamma = \{1, 2, 3, 4\}$. Let $A_1 = \{1, 2\}$, $A_2 = \{1, 2, 3\}$, $A_3 = \{1, 2, 3, 4\}$, and $A_4 = \{2, 3, 4\}$. Let $\mathscr{A} = \{A_\gamma\}_{\gamma \in \Gamma} = \{A_k\}_{k=1}^4$. Then the set $\bigcup_{\gamma \in \Gamma} A_\gamma = \bigcup_{k=1}^4 A_k = \{1, 2, 3, 4\}$. The set $\bigcap_{\gamma \in \Gamma} A_\gamma = \bigcap_{k=1}^4 A_k = \{2\}$.

**Exercises (3.5)** In Exercises 1 through 5, prove the given theorem.

1. Theorem. Let $\mathscr{A} = \{A_\gamma\}_{\gamma \in \Gamma}$ be an indexed collection of sets. Then $(\bigcup_{\gamma \in \Gamma} A_\gamma)^C = \bigcap_{\gamma \in \Gamma} A_\gamma^C$.

2. Theorem. Let $\mathscr{A} = \{A_\gamma\}_{\gamma \in \Gamma}$ be an indexed collection of sets. Then $(\bigcap_{\gamma \in \Gamma} A_\gamma)^C = \bigcup_{\gamma \in \Gamma} A_\gamma^C$.

3. Theorem. Let $A$ be a set, and let $\mathscr{B} = \{B_\delta\}_{\delta \in \Delta}$ be an indexed collection of sets. Then $A \cup (\bigcap_{\delta \in \Delta} B_\delta) = \bigcap_{\delta \in \Delta} (A \cup B_\delta)$.

4. Theorem. Let $A$ be a set, and let $\mathscr{B} = \{B_\delta\}_{\delta \in \Delta}$ be an indexed collection of sets. Then $A \cap (\bigcup_{\delta \in \Delta} B_\delta) = \bigcup_{\delta \in \Delta} (A \cap B_\delta)$.

5. Theorem. For each $k \in \mathbb{N}$, let $A_k$ be a set. Then for each $n \in \mathbb{N}$, $\bigcup_{k=1}^{n+1} A_k = (\bigcup_{k=1}^{n} A_k) \cup A_{n+1}$, and $\bigcap_{k=1}^{n+1} A_k = (\bigcap_{k=1}^{n} A_k) \cap A_{n+1}$.

6. Find $\bigcap_{k \in \mathbb{N}} \{m \in \mathbb{N} \mid m \geq k\}$.

7. Find $\bigcap_{k \in \mathbb{N}} \{m \in \mathbb{N} \mid m \leq k\}$.

**Definition.** Let $A$ be a set, and let $\mathscr{S} = \{S_\gamma\}_{\gamma \in \Gamma}$ be a collection of subsets of $A$. The collection $\mathscr{S}$ is a *partition* of $A$ if the following conditions hold:

a. For all $\gamma \in \Gamma$, $S_\gamma \neq \varnothing$.

b. For all $x \in A$, there exists $\gamma \in \Gamma$ such that $x \in S_\gamma$.

c. For all $\beta, \gamma \in \Gamma$, if $\beta \neq \gamma$ then $S_\beta \cap S_\gamma = \varnothing$.

**Remark.** In the foregoing definition, we have written $\mathscr{S}$ as an indexed set $\{S_\gamma\}_{\gamma \in \Gamma}$. We may also write this definition without reference to an indexing set $\Gamma$. The two definitions are equivalent; only the notation is different. The unindexed version is as follows.

**Definition.** Let $A$ be a set, and let $\mathscr{S}$ be a collection of subsets of $A$. The collection $\mathscr{S}$ is a *partition* of $A$ if the following conditions hold:

a. For all $S \in \mathscr{S}$, $S \neq \varnothing$.

b. For all $x \in A$, there exists $S \in \mathscr{S}$ such that $x \in S$.

c. For all $S, T \in \mathscr{S}$, if $S \neq T$ then $S \cap T = \varnothing$.

**Example (3.13)** Let $A = \{1, 2, 3, 4, 5, 6, 7, 8, 9, 10\}$. Let $\Gamma = \{1, 2, 3, 4\}$. Let $S_1 = \{1, 3, 7\}$; let $S_2 = \{5, 6, 8, 9\}$; let $S_3 = \{4, 10\}$; and let $S_4 = \{2\}$. Then $\mathscr{S} = \{S_\gamma\}_{\gamma \in \Gamma}$ is a partition of $A$.

**Theorem (3.3)** Let $A$ be a nonempty set, and let $\sim$ be an equivalence relation on $A$, and let $x, y \in A$. Let $\sim$ be an equivalence relation on $A$. If $y \in [x]$ then $[y] = [x]$.

**Proof.** Suppose that $y \in [x]$. Since $y \in [x]$, $x \sim y$. Let $z \in [y]$. Since $z \in [y]$, $y \sim z$. Since $x \sim y$ and $y \sim z$ and $\sim$ is transitive, $x \sim z$. Since $x \sim z$, $z \in [x]$.

Thus, for all $z \in [y]$, $z \in [x]$. That is, $[y] \subseteq [x]$.

Let $t \in [x]$. Since $t \in [x]$, $x \sim t$. Since $x \sim t$ and $\sim$ is symmetric, $t \sim x$. Since $t \sim x$ and $x \sim y$, $t \sim y$. Since $t \sim y$ and $\sim$ is symmetric, $y \sim t$. Since $y \sim t$, $t \in [y]$.

Thus, for all $t \in [x]$, $t \in [y]$. That is, $[x] \subseteq [y]$.

Since $[y] \subseteq [x]$ and $[x] \subseteq [y]$, $[x] = [y]$.

Therefore, for each nonempty set $A$, for each equivalence relation $\sim$ on $A$, for all $x$, $y \in A$, if $y \in [x]$ then $[x] = [y]$. Q.E.D.

**Theorem (3.4)** Let $A$ be a nonempty set, and let $\sim$ be an equivalence relation on $A$. Then $A/\sim$ is a partition of $A$.

**Proof.** Let $A$ be a nonempty set, and let $\sim$ be an equivalence relation on $A$. First we will show that for every $S \in A/\sim$, the set $S$ is nonempty. Let $S \in A/\sim$. Since $S \in A/\sim$, there exists $x \in A$ such that $S = [x]$. Since $\sim$ is an equivalence relation, $\sim$ is reflexive. Thus $x \sim x$. Since $x \sim x$, it follows that $x \in [x]$. Since $[x] = S$, $x \in S$. Thus $S \neq \varnothing$. Hence for every $S \in A/\sim$, $S \neq \varnothing$.

Next we will show that for all $x \in A$ there exists $S \in A/\sim$ such that $x \in S$. Let $x \in A$. Since $\sim$ is reflexive, $x \sim x$. Since $x \sim x$, it follows that $x \in [x]$. By definition of $A/\sim$, the equivalence class $[x] \in A/\sim$ . Hence for each $x \in A$ there exists $S \in A/\sim$ (namely, $S = [x]$) such that $x \in S$.

Finally, we will show that for all $S, T \in A/\sim$, if $S \neq T$ then $S \cap T = \varnothing$. By way of contradiction, suppose that there exist $S$, $T \in A/\sim$ such that $S \neq T$ and $S \cap T \neq \varnothing$. Since $S$, $T \in A/\sim$, by definition there exist $x$, $y \in A$ such that $S = [x]$ and $T = [y]$. Since $S \cap T \neq \varnothing$, $[x] \cap [y] \neq \varnothing$. Therefore, there exists $z \in [x] \cap [y]$. Since $z \in [x]$, by Theorem 3.3 $[x] = [z]$. Similarly, since $z \in [y]$, $[y] = [z]$. Since $[x] = [z]$ and $[y] = [z]$, $[x] = [y]$. That is, $S = T$. But by hypothesis $S \neq T$. Hence $S = T$ and $S \neq T$. $\rightarrow\leftarrow$

Our hypothesis has led to a contradiction and is therefore false. Therefore, for all $S, T \in A/\sim$, if $S \neq T$ then $S \cap T = \varnothing$.

Since $A/\sim$ satisfies the three defining conditions for a partition, $A/\sim$ is a partition of $A$.

Therefore, for each set $A$ and each equivalence relation $\sim$ on $A$, the set of equivalence classes $A/\sim$ is a partition of $A$. Q.E.D.

**Theorem (3.5)** Let $A$ be a set, and let $\mathscr{S}$ be a partition of $A$. There is an equivalence relation $\sim$ on $A$ such that $\mathscr{S} = A/\sim$ .

**Proof.** The proof is left as an exercise for the reader.

**Example (3.14)** Let $A = \{1, 2, 3, 4, 5, 6, 7\}$ and let $\mathscr{S} = \{S_k\}_{k=1}^{3}$ where $S_1 = \{1, 2, 7\}$, $S_2 = \{3, 5\}$, and $S_3 = \{4, 6\}$. Then the equivalence relation associated with $\mathscr{S}$ is the relation $\{(1, 1), (2, 2), (7, 7), (1, 2), (1, 7), (2, 7), (2, 1), (7, 1), (7, 2), (3, 3), (5, 5), (3, 5), (5, 3), (4, 4), (6, 6), (4, 6), (6, 4)\}$.

**Remark.** The description of the relation $\sim$ as a set of ordered pairs is not very satisfying. A diagram showing an arrow from $a$ to $b$ whenever $a \sim b$ is slightly easier to interpret, but cluttered and hard to read.

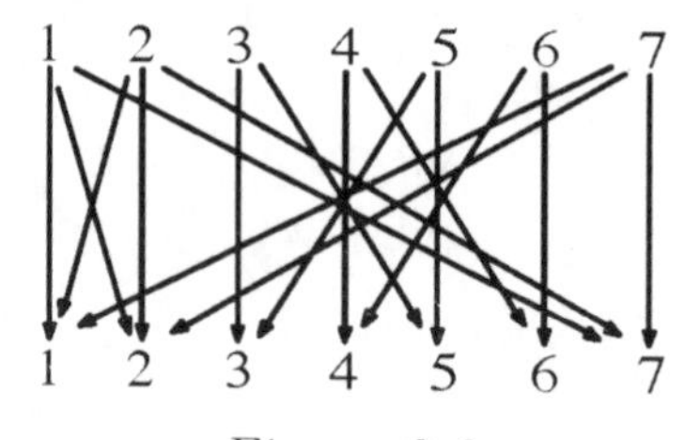

*Figure 3.6*

Since $\sim$ is an equivalence relation, it is easier to diagram it like this:

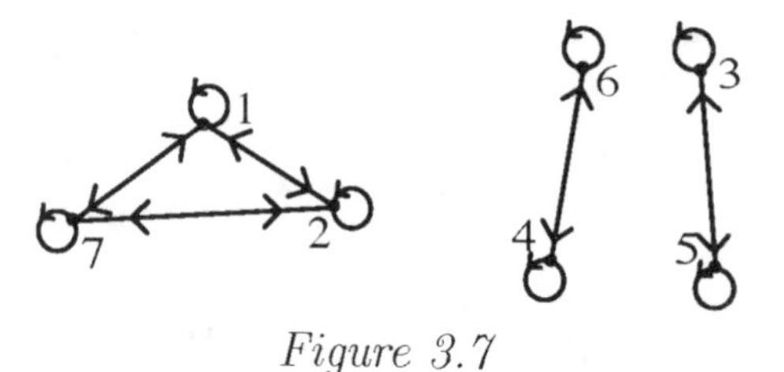

*Figure 3.7*

This amounts to thinking of the equivalence relation $\sim$ in terms of its equivalence classes.

## Exercises (3.6)

1. Prove Theorem 3.5.

**2.** Let $A = \{13, 89, 421, 5, 2, 9, 7\}$ and let $\mathscr{S} = \{S_k\}_{k=4}$ where $S_1 = \{13, 7\}$, $S_2 = \{89, 9, 2\}$, $S_3 = \{421\}$, and $S_4 = \{5\}$. Since $\mathscr{S}$ is a partition of $A$, there is an equivalence relation $\sim$ such that $\mathscr{S} = A/\!\sim$. Explicitly define the relation $\sim$ by listing between set brackets the ordered pairs which belong to $\sim$.

For exercises 3 and 4, let $A = \{1, 2, 3, 4, 5, 6, 7, 8\}$, and let $\sim$ be the relation $\{(x, y) \in A^2 \mid (x-5)^2 = (y-5)^2\}$.

**3.** Show that $\sim$ is an equivalence relation on $A$.

**4.** List the equivalence classes of the relation $\sim$.

**Remark.** Equivalence relations are used extensively in all areas of mathematics. One of the epigraphs to this chapter refers to the fact that the rational numbers may be regarded as equivalence classes under the equivalence relation $\sim$ on $\mathbb{Z} \times \mathbb{N}$ such that $(x, y) \sim (u, v)$ if and only if $xv = uy$. (This was one of a baffling collection of set-theoretical facts that schoolteachers in the 1960s were told to impart to students. This experiment was called New Math. It was a failure, largely because the teachers did not know set theory and so could not possibly teach it to their students.) Order relations are also widely used, so we will discuss them briefly.

**Definition.** Let $A$ be a set, and let $\boldsymbol{r}$ be a relation on $A$. Then $\boldsymbol{r}$ is *antisymmetric* provided that for all $x, y \in A$, if $x\,\boldsymbol{r}\,y$ and $y\,\boldsymbol{r}\,x$ then $x = y$.

**Definition.** Let $A$ be a set, and let $\boldsymbol{r}$ be a relation on $A$. Then $\boldsymbol{r}$ is a *partial order relation* (or *partial order*) if $\boldsymbol{r}$ is reflexive, antisymmetric, and transitive.

**Example (3.15)** The relation $\leq$ is a partial order relation on $\mathbb{R}$.

**Remark.** The quintessential equivalence relation is equality $(=)$, and the quintessential partial order is $\leq$ .

**Exercises (3.7)** For each given relation $r$, determine whether or not $r$ is a partial order. Prove that your answer is correct.

1. Let $\mathscr{S}$ be a collection of sets. Let $r$ be the relation on $\mathscr{S}$ such that $X \, r \, Y$ if and only if $X \subseteq Y$.

2. Let $r$ be the relation $\{(m, n) \in \mathbb{N}^2 \mid m \mid n\}$.

3. Let $r$ be the relation $<$ on $\mathbb{R}$.

4. Let $r$ be the relation $\geq$ on $\mathbb{R}$.

5. Let $r$ be the relation $=$ on $\mathbb{R}$.

6. Let $r$ be the relation on $\mathbb{N}^2$ such that $(x, y) \, r \, (u, v)$ if and only if $x \leq u$ and $y \leq v$.

7. Let $r$ be the relation on $\mathbb{N}^2$ such that $(x, y) \, r \, (u, v)$ if and only if

   (a) $x + y < u + v$; or

   (b) $x + y = u + v$ and

      (i) if $x + y$ is even then $x \geq u$ and $y \leq v$;

      (ii) if $x + y$ is odd then $x \leq u$ and $y \geq v$.

8. Let $r$ be the relation on $\mathbb{N}^2$ such that $(x, y) \, r \, (u, v)$ if and only if $x + y = u + v$.

9. Let $r$ be the relation on $\mathbb{Q}$ such that $x \, r \, y$ if and only if there exists $n \in \mathbb{N}$ such that $nx = y$.

10. Let $r$ be the relation on $\mathbb{N}^2$ such that $(x, y) \, r \, (u, v)$ if and only if (1) $x < u$, or (2) $x = u$ and $y \leq v$.

11. Disprove by counterexample: False Proposition. For each set $A$ and each relation $r$ on $A$, if $r$ is an equivalence relation then $r$ is not a partial order.

**Notation.** Just as $\sim$ is often used to represent an equivalence relation, a partial order is often denoted by the symbol $\preccurlyeq$ .

**Definitions.** Let $A$ be a set, let $x,\ y \in A$, and let $\preccurlyeq$ be a partial order relation on $A$.

**(i)** The elements $x$ and $y$ are *comparable* if $x \preccurlyeq y$ or $y \preccurlyeq x$.

**(ii)** The relation $\preccurlyeq$ is a *total order relation* (or *total order*) on $A$ if, for all $z,\ v \in A$, $z$ and $v$ are comparable.

**Exercises (3.8)** For each relation $r$ in Exercises 3.7, if $r$ is a partial order then determine whether or not $r$ is a total order on the given set. Prove that your answer is correct.

**Definition.** Let $A$ be a set, let $\preccurlyeq$ be a partial order on $A$, and let $\preccurlyeq_*$ be a total order on $A$. The total order $\preccurlyeq_*$ is *compatible* with the partial order $\preccurlyeq$ provided that for each $x,\ y \in A$, if $x \preccurlyeq_* y$, then $x \preccurlyeq y$ or $x$ and $y$ are not comparable.

**Example (3.16)** Consider the relation $\preccurlyeq$ on $\mathbb{N}$ such that $x \preccurlyeq y$ if and only if $x - y$ is even and $x \leq y$. Then $\preccurlyeq$ is a partial order on $\mathbb{N}$. The relation $\preccurlyeq$ is not a total order. The numbers 1 and 2, for example, are not comparable.

There are several total orders that are compatible with $\preccurlyeq$. We offer two examples of such total orders.

1. Let $\preccurlyeq_*$ be the relation on $\mathbb{N}$ such that $x \preccurlyeq y$ if and only if (a) $x - y$ is even and $x \leq y$, or (b) $x$ is odd and $y$ is even.

2. Let $\preccurlyeq_*$ be the relation on $\mathbb{N}$ such that $x \preccurlyeq y$ if and only if $x \leq y$.

Exercises (3.9)

1. Theorem. Let $\preccurlyeq$ be the relation on $\mathbb{N}$ such that $x \preccurlyeq y$ if and only if $x - y$ is even and $x \leq y$. Then $\preccurlyeq$ is a partial order on $\mathbb{N}$.

2. Theorem. Let $\preccurlyeq_*$ be the relation on $\mathbb{N}$ such that $x \preccurlyeq y$ if and only if (a) $x - y$ is even and $x \leq y$, or (b) $x$ is odd and $y$ is even. Then $\preccurlyeq_*$ is a total order, and is compatible with the order $\preccurlyeq$ defined in Exercise 1.

3. Prove that the two total order relations defined in Example 3.16 are not equal.

In Exercises 4 through 7, find a total order relation $\preccurlyeq_*$ that is compatible with the given partial order relation. Prove that your solution is correct. There is more than one correct solution to each problem.

4. For all $x, y \in \mathbb{Z}$, $x \preccurlyeq y$ if and only if $x^2 \leq y^2$ and $xy \geq 0$.

5. For all $x, y, u, v \in \mathbb{N}$, $(x, y) \preccurlyeq (u, v)$ if and only if $x \leq u$ and $y \leq v$.

6. Let $A$ be the set $\{1, 2, 3, 4, 5, 6, 7\}$. Let $\preccurlyeq$ be the partial order on $A$ such that $x \preccurlyeq y$ if and only if $x \mid y$.

7. For all $x, y, u, v \in \mathbb{N}$, $(x, y) \preccurlyeq (u, v)$ if and only if $x \leq u$ and $y \geq v$.

8. Theorem. Let $X, \Gamma$ be sets, and let $\mathscr{A} = \{A_\gamma\}_{\gamma \in \Gamma}$ be a collection of distinct subsets of $X$. Let $\preccurlyeq$ be a partial order relation on $\Gamma$. Let $\preccurlyeq_*$ be the relation on $\mathscr{A}$ such that $A_\alpha \preccurlyeq_* A_\beta$ if $\alpha \preccurlyeq \beta$. Prove the following:

   (a) $\preccurlyeq_*$ is a partial order relation on $\mathscr{A}$.

   (b) If $\preccurlyeq$ is a total order relation on $\Gamma$, then $\preccurlyeq_*$ is a total order relation on $\mathscr{A}$.

# Chapter 4

# Functions

*Nowadays, of course, a function $y = y(x)$ means that there is a class of "arguments" $x$, and to each $x$ there is assigned 1 and only 1 "value" $y$. After some trivial explanations (or none?) we can be balder still that a function is a class $C$ of pairs $(x,y)$ (order within the bracket counting), $C$ being subject (only) to the condition that the $x$'s of different pairs are different.... This clear daylight is now a matter of course, but it replaces an obscurity as of midnight.*[1] *The main step was taken by Dirichlet in 1837 (for functions of a real variable, the argument class consisting of some or all real numbers and the value class confined to real numbers.) The complete emancipation of e.g. propositional functions belongs to the 1920's.*

[1] *The trouble was, of course, an obstinate feeling at the back of the mind that the value of a function "ought" to be got from the argument by "a series of operations."*

*J. E. Littlewood,* Littlewood's Miscellany

*"I beg your pardon," she said. "I haven't seen you for thirty years.* Thirty years. *We're twice as old as we used to be...."*

*The math confused me. Weren't you always twice as old as you used to be?*

*Elizabeth McCracken,* Niagara Falls All Over Again.

**Definition.** Let $A$ and $B$ be sets, and let $f$ be a relation from $A$ to $B$. The relation $f$ is a *function* from $A$ to $B$ if: (1) for every $x \in A$, there exists $y \in B$ such that $(x, y) \in f$; and (2) for all $y_1, y_2 \in B$, if $(x, y_1) \in f$ and $(x, y_2) \in f$, then $y_1 = y_2$.

**Remark.** Conditions (1) and (2) above may be stated more succinctly as follows: for every $x \in A$, there exists a unique $y \in B$ such that $(x, y) \in f$.

**Notation.**

1. Let $A$ and $B$ be sets. The sentence: "Let $f$ be a function from $A$ to $B$," may be written, "Let $f: A \to B$ be a function," or even simply, "Let $f: A \to B$." All three formulations are pronounced, "Let $f$ be a function from $A$ to $B$." A *function on* $A$ is a function from $A$ to $A$.

2. Let $f: A \to B$ be a function, and let $x \in A$, $y \in B$ such that $(x, y) \in f$. We very seldom write, "$(x, y) \in f$." We usually express this idea by writing, "$f(x) = y$." The symbol $f(x)$ is pronounced, "$f$ of $x$."

3. We recall the following definitions from Chapter 3 (with "function" substituted for "relation").

   **Definitions.** Let $A$ and $B$ be sets, and let $f: A \to B$ be a function.

   **(i)** The *domain* of $f$, denoted by Dom $f$, is the set $\{x \in A \mid$ there exists $y \in B$ such that $f(x) = y\}$.

   **(ii)** The *image* of $f$, denoted by Im $f$, is the set $\{x \in B \mid$ there exists $x \in A$ such that $f(x) = y\}$. The image of $f$ is also called the *range* of $f$.

Notice that the definition of *function* implies that $A =$ Dom $f$. The set $B$ is called the *target*, or *co-domain*, of $f$. Notice that Im $f \subseteq B$.

**Remark.** As our notation suggests, we do not usually think of a function as a set of ordered pairs. Instead, we often think of a function as a rule of assignment, as shown below.

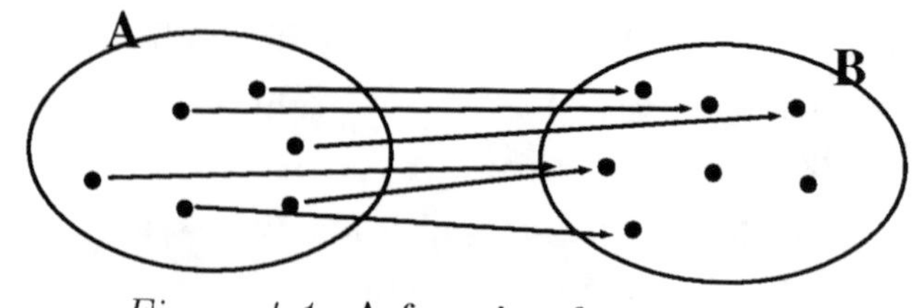


*Figure 4.1:* A function from $A$ to $B$

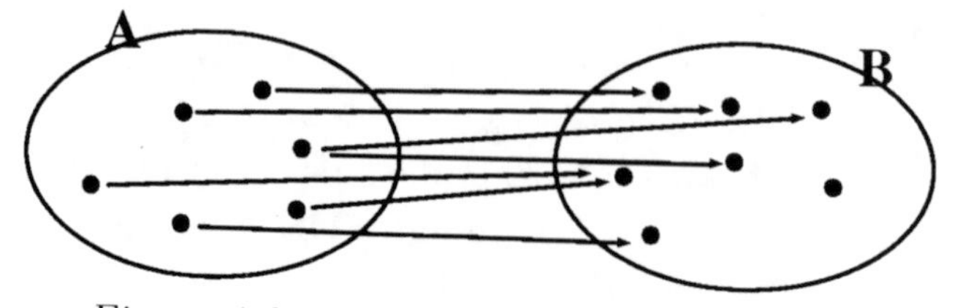


*Figure 4.2*: Not a function from $A$ to $B$

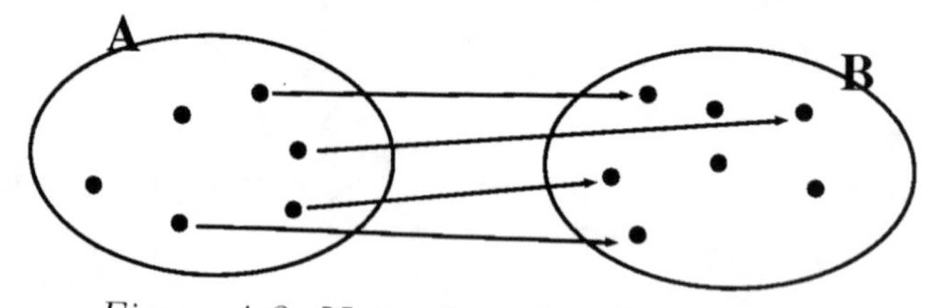


*Figure 4.3*: Not a function from $A$ to $B$

Some people like to think of a function $f\colon A \to B$ as a machine that takes in an element $x$ of $A$ and puts out an element $f(x)$ of $B$.

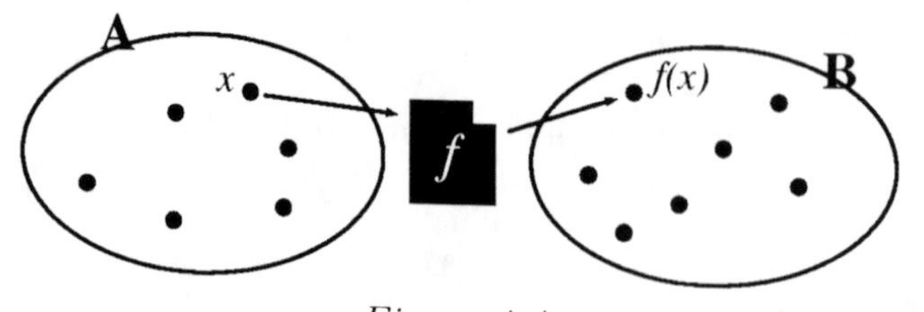


*Figure 4.4*

When we think of a function $f$ as a set of ordered pairs, we are thinking of the *graph* of $f$.

**Example (4.1)** Consider the function $f: \mathbb{R} \to \mathbb{R}$ defined by $f(x) = x^2 + 1$.

For any $(x, y) \in \mathbb{R}^2$, the point $(x, y)$ is on the graph of $y = x^2 + 1$ if and only if $y = x^2 + 1$. That is, for each $(x, y) \in \mathbb{R}^2$, the point $(x, y)$ is on the graph of $f$ if and only if $f(x) = y$.

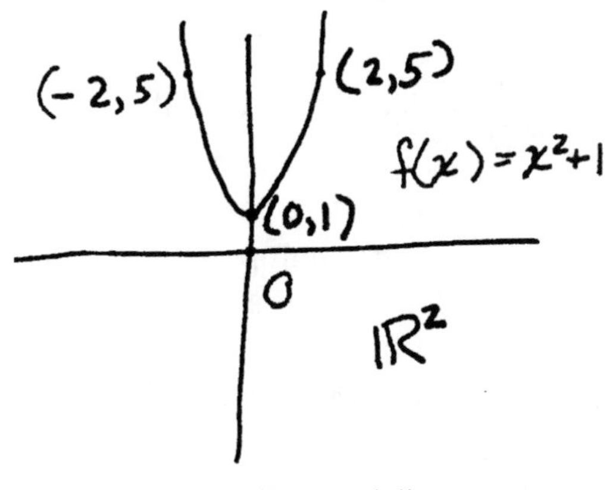


*Figure 4.5*

We may also think of $f$ as a rule of assignment.

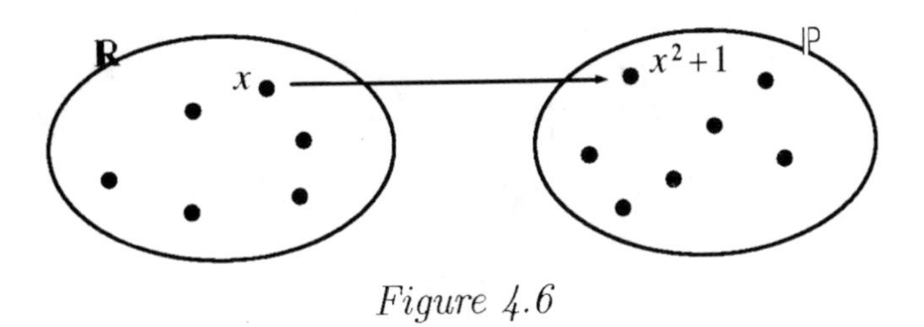


*Figure 4.6*

When the domain of $f$ is the set of real numbers $\mathbb{R}$, it seems natural to draw the graph of $f$. (We need not draw the graph very accurately. The hand-drawn graph of Figure 4.5 is good enough to give us some idea of the function $f$.) This is because $\mathbb{R}$ is the set of all the points on the number line. The real line $\mathbb{R}$ is a continuum, not a set of disconnected points.

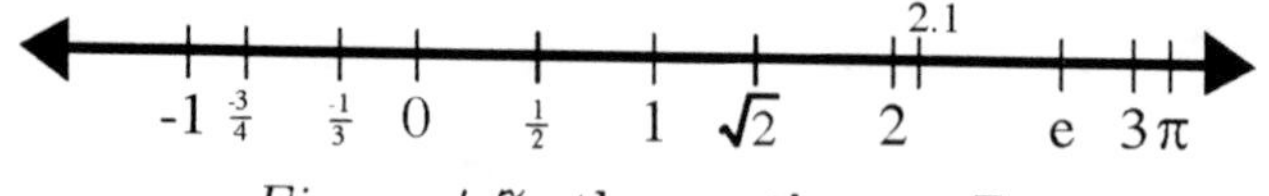

*Figure 4.7*: the continuum $\mathbb{R}$

When the domain of $f$ is $\mathbb{N}$ or a finite set, it is still possible, although less customary, to represent $f$ as a graph. For example, let $A = \{1, 2, 3, 4, 5\}$, and let $f: A \to A$ be the function $f(x) = |x - 2| + 1$. Then $f(1) = 2$, $f(2) = 1$, $f(3) = 2$, $f(4) = 3$, and $f(5) = 4$. That is, $f = \{(1, 2), (2, 1), (3, 2), (4, 3), (5, 4)\}$.

The graph of $f$ can be drawn as shown below.

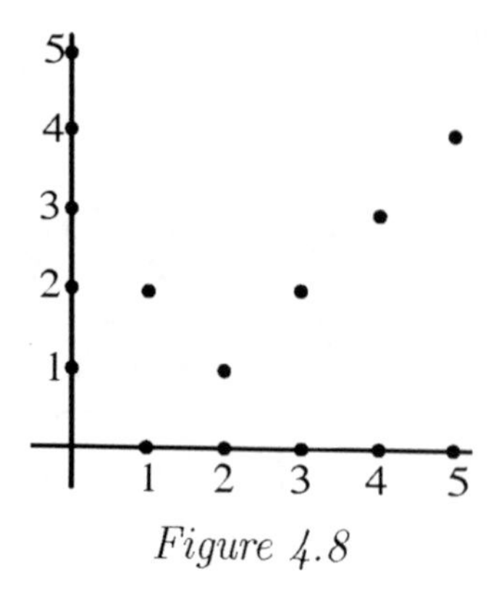

*Figure 4.8*

But we may find it more convenient to think of $f$ this way:

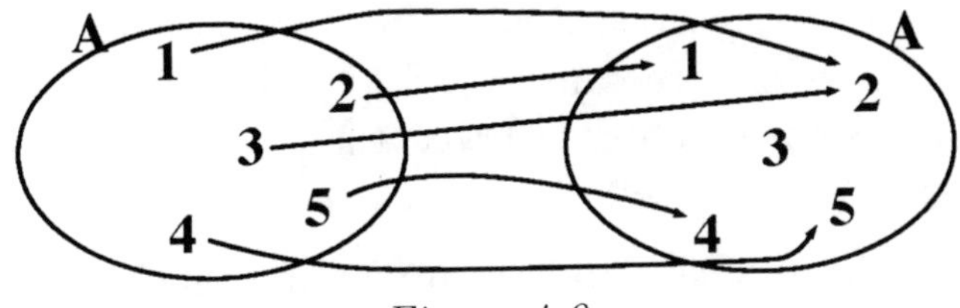


*Figure 4.9*

or as follows:

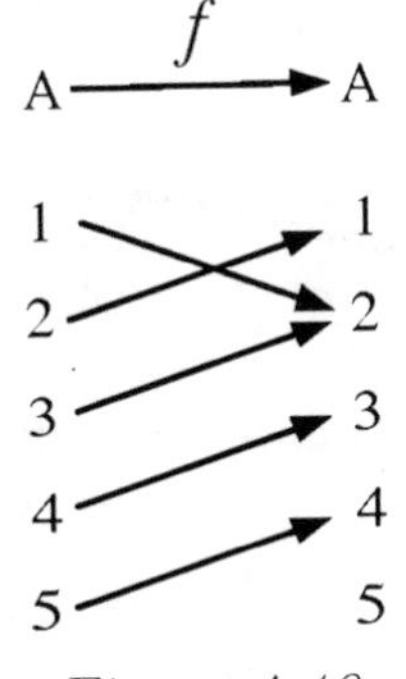


*Figure 4.10*

or as a sequence of equations:

$$f(1) = 2.$$
$$f(2) = 1.$$
$$f(3) = 2.$$
$$f(4) = 3.$$
$$f(5) = 4.$$

All of these ways of thinking of functions are legitimate, and mathematicians make use of all of them.

**Remark.** When students are first introduced to the modern definition of function, many feel unsatisfied with it. For example, let $A = \{1, 2, 3, 4, 5\}$, and let $f: A \rightarrow A$ be the function $f(1) = 4$, $f(2) = 3$, $f(3) = 5$, $f(4) = 2$, $f(5) = 1$. Students often feel that there ought to be a formula for $f$. In fact, this is not necessary.

Historically, the student's feelings are entirely justified. Euler introduced the notation $f(x)$, but he thought that a function should be defined by a formula or a curve drawn freehand, thus:

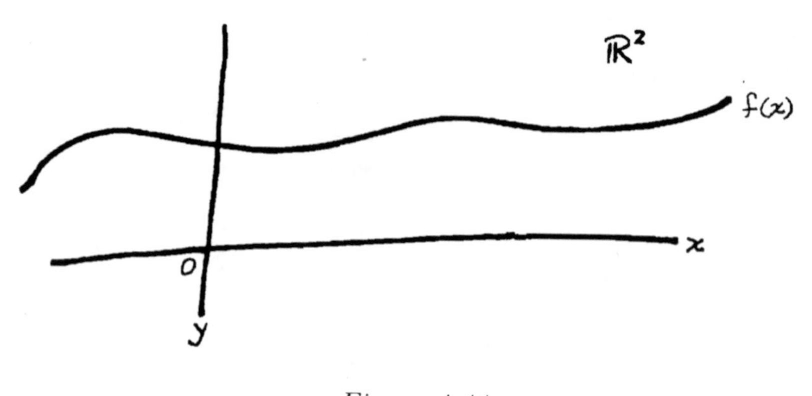


*Figure 4.11*

For a long time, the notion of "function" was very unclear. Georg Cantor wrote our modern definition as part of his set theory.

Thus, the student who at first feels uneasy with the modern definition of function is entirely in tune with the historical feelings of mathematicians. But we are lucky to have such a clear, simple definition, and lucky to have the notation $f(x)$.

## Exercises (4.1)

1. Determine whether or not each relation $r$ is a function from $\{0, 1, 2\}$ to $\{0, 1, 2\}$.

(a) $r = \{(0, 0), (0, 1), (0, 2)\}$.
(b) $r = \{(0, 0), (1, 0), (2, 0)\}$.
(c) $r = \{(1, 0), (2, 1), (1, 1)\}$.
(d) $r = \{(0, 1), (1, 2)\}$.
(e) $r = \{(0, 2), (1, 2), (2, 1), (2, 2)\}$.
(f) $r = \{(1, 2), (2, 1), (0, 0)\}$.

**2.** Determine whether or not each relation given below is a function from $\mathbb{R}$ to $\mathbb{R}$.

(a) $\{(x, y) \in \mathbb{R}^2 \mid x = y\}$
(b) $\{(x, y) \in \mathbb{R}^2 \mid x < y\}$
(c) $\{(x, y) \in \mathbb{R}^2 \mid x^2 = y\}$
(d) $\{(x, y) \in \mathbb{R}^2 \mid x = y^2\}$
(e) $\{(x, y) \in \mathbb{R}^2 \mid x = 2y - 1\}$
(f) $\{(x, y) \in \mathbb{R}^2 \mid y = 1/x\}$
(g) $\{(x, y) \in \mathbb{R}^2 \mid y = x \text{ or } y = -x\}$

**3.** List all functions from $\{1, 2, 3\}$ to $\{1, 2\}$.

**4.** List all functions from $\{1, 2\}$ to $\{1, 2, 3\}$.

The connectives $\neg$, $\vee$, $\wedge$, $\rightarrow$, and $\leftrightarrow$ may be regarded as functions. Let $A = \{\text{T}, \text{F}\}$. The connective $\neg: A \rightarrow A$ is the function defined so that $\neg(\text{T}) = \text{F}$ and $\neg(\text{F}) = \text{T}$. The connective $\vee: A^2 \rightarrow A$ is the function defined so that $\vee((\text{T}, \text{T})) = \text{T}$, $\vee((\text{T}, \text{F})) = \text{T}$, $\vee((\text{F}, \text{T})) = \text{T}$, and $\vee((\text{F}, \text{F})) = \text{F}$.

**5.** Let $A = \{\text{T}, \text{F}\}$. Define the connective $\rightarrow$ as a function from $A^2$ to $A$.

**6.** Let $A = \{\text{T}, \text{F}\}$. Define the connective $\wedge$ as a function from $A^2$ to $A$.

**Definition.** Let $A$ and $B$ be sets, let $C \subseteq A$, and let $f: A \to B$ be a function. The *image of $C$ under $f$*, denoted by f(C), is the set $\{y \in B \mid \text{there exists } x \in C \text{ such that } f(x) = y\}$.

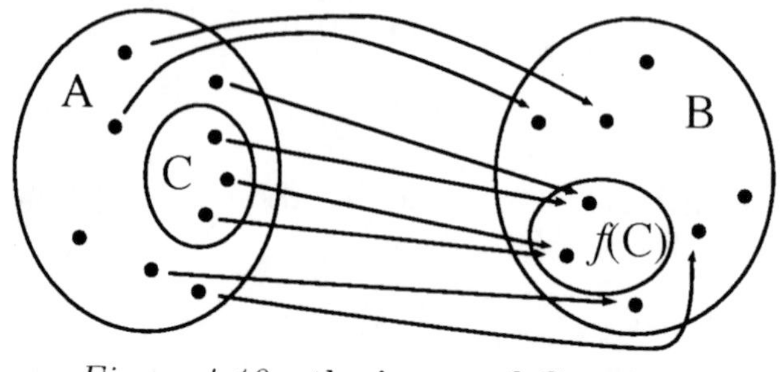


*Figure 4.12:* the image of $C$ under $f$

**Definition.** Let $A$ and $B$ be sets, let $D \subseteq B$, and let $f: A \to B$ be a function. The *inverse image*, or *pre-image*, of $D$ under $f$, denoted by $f^{-1}(D)$, is the set $\{x \in A \mid f(x) \in D\}$.

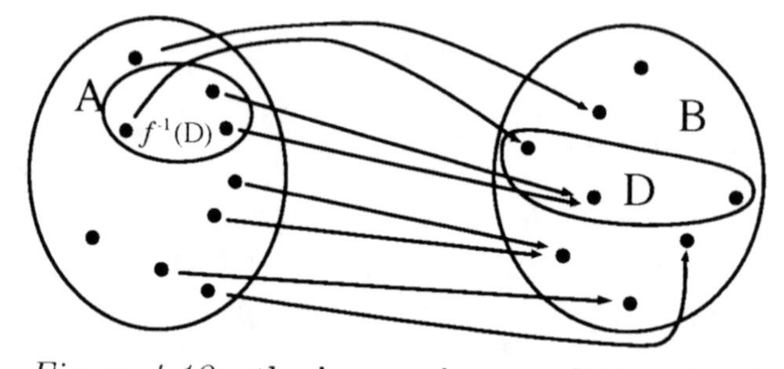


*Figure 4.13:* the inverse image of $D$ under $f$

**Notation.** Let $A$ and $B$ be sets, let $f: A \to B$ be a function, and let $x \in B$. Then $f^{-1}(\{x\})$ denotes the inverse image of the set $\{x\}$ under $f$. Notice that $f^{-1}$ is not the name of a function, and that $f^{-1}(\{x\})$ is a subset of $A$, not an element of $A$.

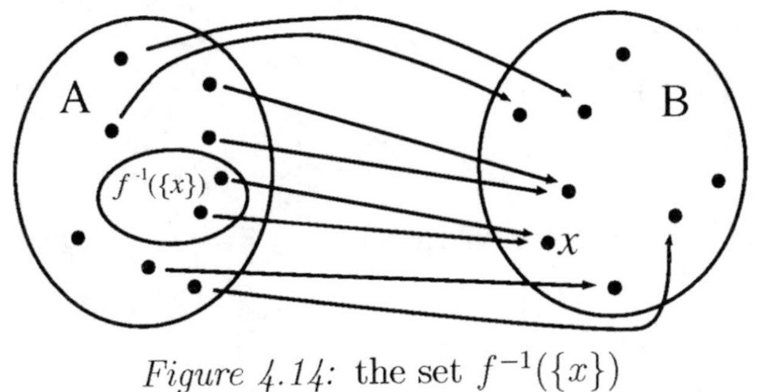


*Figure 4.14:* the set $f^{-1}(\{x\})$

**Example (4.2)** Let $A = \{0, 1, 2, 3, 4, 5\}$ and let $B = \{0, 1, 2, 3, 4, 5, 6\}$. Let $f: A \to B$ be the function $\{(0, 6), (1, 1), (2, 4), (3, 4), (4, 1), (5, 5)\}$.

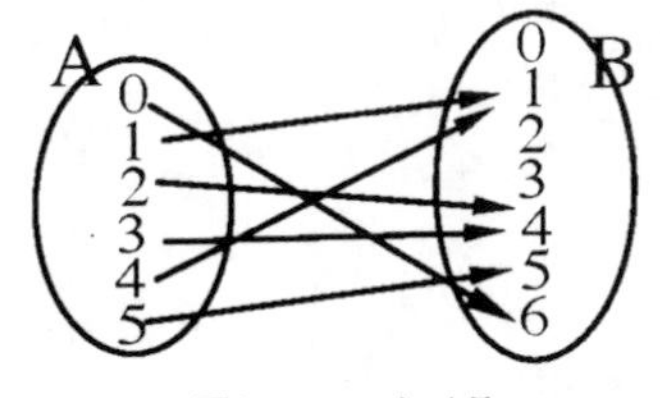


*Figure 4.15*

Then $f(0) = 6$; $f(1) = 1$; $f(2) = 4$; $f(3) = 4$; $f(4) = 1$; and $f(5) = 5$.

Images of sets under $f$ include: $f(\varnothing) = \varnothing$; $f(\{0, 1\}) = \{6, 1\}$; and $f(A) = \{1, 4, 5, 6\}$.

Inverse images of sets under $f$ include: $f^{-1}(\{0\}) = \varnothing$; $f^{-1}(\{1\}) = \{1, 4\}$; $f^{-1}(\{2\}) = \varnothing$; $f^{-1}(\{3\}) = \varnothing$; $f^{-1}(\{4\}) = \{2, 3\}$; $f^{-1}(\{5\}) = \{5\}$; and $f^{-1}(\{6\}) = \{0\}$.

Thus, $f(f^{-1}(\{2\})) = \varnothing$; $f^{-1}(f(\{2\})) = \{2, 3\}$.

**Exercises (4.2)**

1. Let $f: \{0, 1, 2, 3\} \to \{0, 1, 2, 3\}$ be defined by $f = \{(0, 0), (1, 2), (2, 0), (3, 2)\}$. Find $f(\{0, 1\})$, $f^{-1}(\{0\})$, $f(f^{-1}(\{0\})$, and $f^{-1}(f(\{0\}))$.

2. Let $f: \mathbb{R} \to \mathbb{R}$ be defined by $f(x) = 2x - 1$. Let $S = \{x \in \mathbb{R} \mid -1 \leq x < 2\}$.
   (a) Find $f(S)$.
   (b) Find $f^{-1}(S)$.

3. Let $f: \mathbb{R} \to \mathbb{R}$ be defined by $f(x) = 5 - 4x$. Let $S = \{x \in \mathbb{R} \mid -1 \leq x < 2\}$.
   (a) Find $f(S)$.
   (b) Find $f^{-1}(S)$.

4. Let $g: \mathbb{R} \to \mathbb{R}$ be defined by $g(x) = x^2 - 1$. Let $S = \{x \in \mathbb{R} \mid -2 \leq x < 2\}$.
   (a) Find $g(S)$.
   (b) Find $g^{-1}(S)$.
   (c) Find $g(g^{-1}(S))$.
   (d) Find $g^{-1}(g(S))$.

5. Let $g: \mathbb{Z} \to \mathbb{Z}$ be defined by $g(x) = \begin{cases} x + 10 \text{ if } x \text{ is odd;} \\ x - 11 \text{ if } x \text{ is even.} \end{cases}$

   Let $S = \{x \in \mathbb{N} \mid 10 \leq x < 20\}$.

   (a) Find $g(S)$.
   (b) Find $g^{-1}(S)$.
   (c) Find $g(g^{-1}(S))$.
   (d) Find $g^{-1}(g(S))$.

6. Let $f: \mathbb{Z} \to \mathbb{N}$ be defined by $f(x) = \begin{cases} \sqrt{x} \text{ if x is a perfect square;} \\ x^2 \text{ otherwise.} \end{cases}$

   Let $S = \{x \in \mathbb{N} \mid x \leq 10\}$.

   (a) Find $f(S)$.

(**b**) Find $f^{-1}(S)$.
(**c**) Find $f(f^{-1}(S))$.
(**d**) Find $f^{-1}(f(S))$.

**7.** Let $A = \{\text{T}, \text{F}\}$ and let $f: A^2 \to A$ be the connective $\to$ .

(**a**) Find $f((\text{T}, \text{T}))$.
(**b**) Find $f^{-1}(\{f((\text{T}, \text{T}))\})$.
(**c**) Find $f(\{(\text{T}, \text{T}), (\text{T}, \text{F})\})$.
(**d**) Find $f^{-1}(f(\{(\text{T}, \text{T}), (\text{T}, \text{F})\}))$.
(**e**) Find $f^{-1}(\{\text{T}\})$.
(**f**) Find $f(f^{-1}(\{\text{T}\}))$.

In exercises 8 through 21, prove each theorem, and disprove each false proposition by giving a counterexample.

**8.** Theorem. Let $X$ and $Y$ be sets. Let $f: X \to Y$ be a function, and let $A$ and $B$ be subsets of $X$. If $A \subseteq B$, then $f(A) \subseteq f(B)$.

**9.** Theorem. Let $X$ and $Y$ be sets. Let $f: X \to Y$ be a function, and let $A$ and $B$ be subsets of $X$. Then $f(A \cap B) \subseteq f(A) \cap f(B)$.

**10.** False Proposition. Let $X$ and $Y$ be sets. Let $f: X \to Y$, and let $A$ and $B$ be subsets of $X$. Then $f(A \cap B) = f(A) \cap f(B)$.

**11.** Theorem. Let $X$ and $Y$ be sets. Let $f: X \to Y$, and let $A$ and $B$ be subsets of $X$. Then $f(A \cup B) = f(A) \cup f(B)$.

**12.** Theorem. Let $X$ and $Y$ be sets. Let $f: X \to Y$, and let $A$ and $B$ be subsets of $X$. Then $f(A) - f(B) \subseteq f(A - B)$.

**13.** False Proposition. Let $X$ and $Y$ be sets. Let $f: X \to Y$, and let $A$ and $B$ be subsets of $X$. Then $f(A) - f(B) = f(A - B)$.

**14.** Theorem. Let $X$ and $Y$ be sets. Let $f: X \to Y$, and let $A$ and $B$ be subsets of $Y$. If $A \subseteq B$, then $f^{-1}(A) \subseteq f^{-1}(B)$.

15. Theorem. Let $X$ and $Y$ be sets. Let $f: X \to Y$, and let $A$ and $B$ be subsets of $Y$. Then $f^{-1}(A \cap B) = f^{-1}(A) \cap f^{-1}(B)$.

16. Theorem. Let $X$ and $Y$ be sets. Let $f: X \to Y$, and let $A$ and $B$ be subsets of $Y$. Then $f^{-1}(A \cup B) = f^{-1}(A) \cup f^{-1}(B)$.

17. Theorem. Let $X$ and $Y$ be sets. Let $f: X \to Y$, and let $A$ and $B$ be subsets of $Y$. Then $f^{-1}(A - B) = f^{-1}(A) - f^{-1}(B)$.

18. Theorem. Let $X$ and $Y$ be sets. Let $f: X \to Y$, and let $A$ be a subset of $Y$. Then $f(f^{-1}(A)) \subseteq A$.

19. False Proposition. Let $X$ and $Y$ be sets. Let $f: X \to Y$, and let $A$ be a subset of $Y$. Then $f(f^{-1}(A)) = A$.

20. Theorem. Let $X$ and $Y$ be sets. Let $f: X \to Y$, and let $A$ be a subset of $X$. Then $A \subseteq f^{-1}(f(A))$.

21. False Proposition. Let $X$ and $Y$ be sets. Let $f: X \to Y$, and let $A$ be a subset of $X$. Then $A = f^{-1}(f(A))$.

**Definition.** Let $A$, $B$, and $C$ be sets, and let $f: A \to B$ and $g: B \to C$ be functions. The *composition of $g$ with $f$*, denoted by $g \circ f$, is the function $h: A \to C$ defined by $h(x) = g(f(x))$.

**Vocabulary.** The function $g \circ f$ is sometimes called *$g$ composed with $f$*.

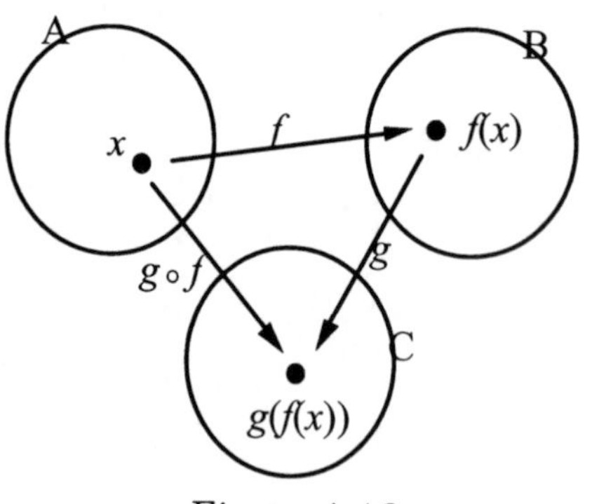


*Figure 4.16*

**Remarks.** We sometimes represent the idea that for all $x \in A$, $(g \circ f)(x) = g(f(x))$ by means of a *commutative diagram*, thus:

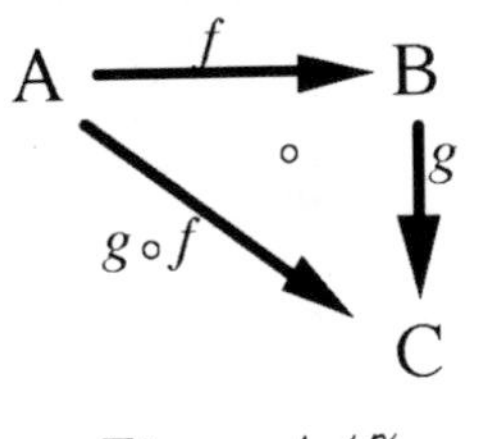


*Figure 4.17*

Let $f: A \to B$ and let $g: B \to C$. It can be hard to remember that for each $x \in A$, $(g \circ f)(x)$ is found by first applying $f$ to $x$ and then applying $g$ to $f(x)$. It can be very helpful to draw a diagram with arrows.

**Theorem (4.1)** Let $A$, $B$, $C$ and $D$ be sets, and let $f: A \to B$ and $g: C \to D$. Then $f = g$ if and only if $A = C$ and for all $x \in A$, $f(x) = g(x)$.

**Proof.** Suppose that $f = g$. Let $x \in A$. Since $A = \text{Dom } f$, there exists $y \in B$ such that $(x, y) \in f$. Since $f = g$ and $(x, y) \in f$, $(x, y) \in g$. Since $(x, y) \in g$, and $C = \text{Dom } g$, it follows that $x \in C$.

Since $(x, y) \in f$, $y = f(x)$. Since $(x, y) \in g$, $y = g(x)$. Thus $f(x) = g(x)$.

Thus, for all $x \in A$, $x \in C$ and $f(x) = g(x)$. That is, $A \subseteq C$ and for all $x \in A$, $f(x) = g(x)$.

Let $z \in C$. Since $C = \text{Dom } g$, there exists $v \in D$ such that $(z, v) \in g$. Since $(z, v) \in g$ and $f = g$, $(z, v) \in f$. Since $(z, v) \in f$ and $A = \text{Dom } f$, $z \in A$.

Thus, for all $x \in C$, $x \in A$. That is, $C \subseteq A$. Since $A \subseteq C$ and $C \subseteq A$, $A = C$.

Therefore, if $f = g$, then $A = C$ and for all $x \in A$, $f(x) = g(x)$.

It remains to show that if $A = C$ and for all $x \in A$, $f(x) = g(x)$, then $f = g$.

Suppose that $A = C$ and that for all $x \in A$, $f(x) = g(x)$. Let $(z, y) \in f$. Since $(z, y) \in f$ and $A = \text{Dom } f$, it follows that $z \in A$. Since $z \in A$, by hypothesis $f(z) = g(z)$. Since $y = f(z)$, $y = g(z)$. Since $y = g(z)$, $(z, y) \in g$.

Thus, for all $(z, y) \in f$, $(z, y) \in g$. That is, $f \subseteq g$. It remains to show that $g \subseteq f$.

Let $(u, v) \in g$. Since $(u, v) \in g$ and $C = \text{Dom } g$, $u \in C$. Since $u \in C$ and $A = C$, $u \in A$. Since $u \in A$, by hypothesis $f(u) = g(u)$. Since $u \in A$ and $g(u) = v$, $f(u) = v$. Since $f(u) = v$, $(u, v) \in f$.

Thus, for all $(u, v) \in g$, $(u, v) \in f$. That is, $g \subseteq f$.

Since $f \subseteq g$ and $g \subseteq f$, it follows that $f = g$.

Hence, if $A = C$ and for all $x \in A$, $f(x) = g(x)$, then $f = g$.

Therefore, for all sets $A$, $B$, $C$, $D$ and all functions $f: A \to B$ and $g: C \to D$, $f = g$ if and only if $A = C$ and for all $x \in A$, $f(x) = g(x)$. Q.E.D.

**Remarks.** Theorem 4.1 states that the functions $f: A \to B$ and $g: C \to D$ are the same if and only if $A = C$ and for all $x \in A$, $f(x) = g(x)$. Notice that the targets $B$ and $C$ need not be identical. (See Exercise 4.3.6.)

When mathematicians prove that two functions $f$ and $g$ are equal, we usually use Theorem 4.1 rather than prove that $f \subseteq g$ and $g \subseteq f$. Thus, to prove that $f = g$, we usually prove that $\text{Dom } f = \text{Dom } g$ and for all $x \in \text{Dom } f$, $f(x) = g(x)$. The next proof illustrates this method.

**Theorem (4.2)** Let $A$, $B$, $C$ and $D$ be sets, and let $f: A \to B$, $g: B \to C$, and $h: C \to D$. Then $(h \circ g) \circ f = h \circ (g \circ f)$.

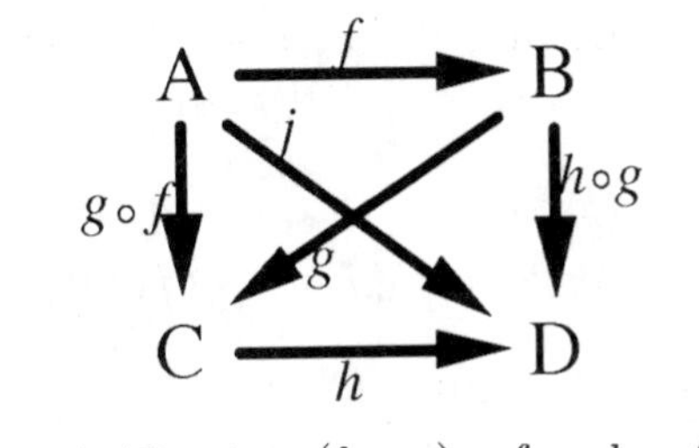


*Figure 4.18:* $j = (h \circ g) \circ f = h \circ (g \circ f)$

**Proof.** The domain of $(h \circ g) \circ f$ is $A$, and the domain of $h \circ (g \circ f)$ is $A$. Hence Dom $(h \circ g) \circ f$ = Dom $h \circ (g \circ f)$. Let $x \in A$. Then $((h \circ g) \circ f)(x) = (h \circ g)(f(x)) = h(g(f(x)))$, and $(h \circ (g \circ f))(x) = h((g \circ f)(x))$ $= h(g(f(x)))$. Since Dom $(h \circ g) \circ f$ = Dom $h \circ (g \circ f)$ and for all $x \in A$, $((h \circ g) \circ f)(x) = (h \circ (g \circ f))(x)$, it follows that $(h \circ g) \circ f$ $= h \circ (g \circ f)$. Q.E.D.

**Exercises (4.3)**

1. Let $f: \mathbb{Z} \to \mathbb{N}$ be defined by $f(x) = \begin{cases} \sqrt{x} \text{ if x is a perfect square;} \\ x^2 \text{ otherwise.} \end{cases}$

   Let $g: \mathbb{N} \to \mathbb{Z}$ be defined by $g(x) = \begin{cases} x + 10 \text{ if } x \text{ is odd;} \\ x - 11 \text{ if } x \text{ is even.} \end{cases}$

   Find $(g \circ f)(2)$, $(g \circ f)(5)$, $(g \circ f)(9)$, and $(g \circ f)(16)$.

2. Let $f: \mathbb{R} \to \mathbb{R}$ be defined by $f(x) = 2x - 1$. Let $g: \mathbb{R} \to \mathbb{R}$ be defined by $g(x) = \frac{1}{2} + \frac{x}{2}$. Find and simplify the functions $(g \circ f)(x)$, $(f \circ g)(x)$, $(f \circ f)(x)$, and $(g \circ g)(x)$.

In Exercises 3 through 6, prove the theorems and disprove the false proposition by giving a counterexample.

3. Theorem. Let $A$, $B$, and $C$ be sets. Let $f: A \to B$ and $g: B \to C$. Then Dom $g \circ f$ = Dom $f$ and Im $g \circ f \subseteq$ Im $g$.

4. <u>Theorem.</u> Let $A$, $B$, and $C$ be sets. Let $f: A \to B$ and $g: B \to C$, and let $D \subseteq C$. Then $(g \circ f)^{-1}(D) = f^{-1}(g^{-1}(D))$.

5. <u>Theorem.</u> Let $A$, $B$, and $C$ be sets. Let $f: A \to B$ and $g: B \to C$, and let $D \subseteq A$. Then $(g \circ f)(D) = g(f(D))$.

6. <u>False Proposition.</u> For all sets $A$, $B$, $C$, $D$ and for all functions $f: A \to B$ and $g: C \to D$, if $f = g$ then $B = D$.

**<u>Definition.</u>** Let $A$ be a set. The *identity function* on $A$ is the function $f: A \to A$ defined by $f(x) = x$. The symbol $I_A$ denotes the identity function on $A$.

**<u>Theorem</u> (4.3)** Let $A$ and $B$ be sets, and let $f: A \to B$. Then $f \circ I_A = f$ and $I_B \circ f = f$.

**<u>Proof.</u>** Exercise.

**<u>Definition.</u>** Let $A$ and $B$ be sets and let $f: A \to B$. A function $g: B \to A$ is an *inverse* (or *inverse function*) of $f$ if $g \circ f = I_A$ and $f \circ g = I_B$. If there exists $g: B \to A$ such that $g$ is an inverse of $f$, then $f$ is *invertible.*

**<u>Theorem</u> (4.4)** Let $A$ and $B$ be sets, and let $f: A \to B$. If $f$ is invertible then the inverse of $f$ is unique.

**<u>Proof.</u>** Suppose that $f$ is invertible, and that $g: B \to A$ and $h: B \to A$ are inverses of $f$. We will show that $g = h$.

By Theorem 4.2, $(g \circ f) \circ h = g \circ (f \circ h)$. Since $g$ is an inverse of $f$, $g \circ f = I_A$. By Theorem 4.3, $I_A \circ h = h$. Thus $(g \circ f) \circ h = I_A \circ h = h$.

Since $h$ is an inverse of $f$, $f \circ h = I_B$. By Theorem 4.3, $g \circ I_B = g$. Thus $g \circ (f \circ h) = g \circ I_B = g$.

Thus $h = (g \circ f) \circ h = g \circ (f \circ h) = g$.

Therefore, if $f: A \to B$ is invertible, then the inverse of $f$ is unique. Q.E.D.

**Theorem (4.5)** Let $A$ and $B$ be sets, and let $f: A \to B$, $g: B \to A$ be functions such that $g$ is the inverse of $f$. Let $D \subseteq B$. Then $f^{-1}(D) = g(D)$.

**Proof.** Recall that $f^{-1}(D)$ is the inverse image of the set $D$ under $f$. Thus $f^{-1}(D) = \{x \in A \mid f(x) \in D\}$. Similarly, $g(D)$ is the image of the set $D$ under $g$. That is, $g(D) = \{x \in A \mid$ there exists $y \in D$ such that $g(y) = x\}$

Let $a \in f^{-1}(D)$. Then $a \in A$ and $f(a) \in D$. Since $g$ is the inverse function of $f$, $g(f(a)) = I_A(a) = a$. Let $b = f(a)$. Then $b \in D$ and $g(b) = a$. Hence $a \in g(D)$. Thus for all $a \in f^{-1}(D)$, $a \in g(D)$. That is, $f^{-1}(D) \subseteq g(D)$.

Let $a \in g(D)$. Then there exists $b \in D$ such that $g(b) = a$. Since $g$ is the inverse function of $f$, $f(g(b)) = I_B(b) = b$. Thus $f(a) = b$. Since $b \in D$ and $f(a) = b$, $a \in f^{-1}(D)$. Thus for all $a \in g(D)$, $a \in f^{-1}(D)$. That is, $g(D) \subseteq f^{-1}(D)$.

Thus, $g(D) = f^{-1}(D)$.

Therefore, for all sets $A$, $B$, $D$, for all functions $f: A \to B$ and $g: B \to A$, if $g$ is the inverse function of $f$ and $D \subseteq B$, then $f^{-1}(D) = g(D)$. Q.E.D.

**Notation.** Let $A$ and $B$ be sets, let $f: A \to B$ and $g: B \to A$ such that $g$ is the inverse function of $f$, and let $D \subseteq A$. The function $g$ is usually denoted by $f^{-1}: B \to A$. (When used as the name of a function, the symbol $f^{-1}$ is pronounced "$f$ inverse.") Theorem 4.5 shows that the use of the symbol $f^{-1}(D)$ to denote the inverse image of $D$ under $f$ does not conflict with the use of $f^{-1}(D)$ to denote the image of $D$ under the function $f^{-1}$.

**Exercises (4.4)** For each function $f: A \to B$, find the inverse function $f^{-1}: B \to A$. Show that $f^{-1} \circ f = I_A$ and $f \circ f^{-1} = I_B$.

1. $f: \mathbb{R} \to \mathbb{R}$ defined by $f(x) = 1 - x$.

2. $f: \mathbb{Z} \to \mathbb{Z}$ defined by $f(x) = x + 10$.

3. $f: \mathbb{N} \to \mathbb{Z}$ defined by $f(x) = \begin{cases} \frac{x-1}{2}, & \text{if } x \text{ is odd;} \\ \frac{-x}{2}, & \text{if } x \text{ is even.} \end{cases}$

4. Disprove the following false proposition by giving a counterexample.

   False Proposition. Let $A$ and $B$ be sets, let $f: A \to B$ and let $g: B \to A$. If $g \circ f = I_A$, then $g$ is the inverse function of $A$.

**Definition.** Let $A$ and $B$ be sets, and let $f: A \to B$. The function $f$ is *one-to-one* (or *injective*) if for all $x$, $y \in A$, if $f(x) = f(y)$ then $x = y$. A one-to-one function is called an *injection*. The phrase "one-to-one" may be abbreviated "1-1."

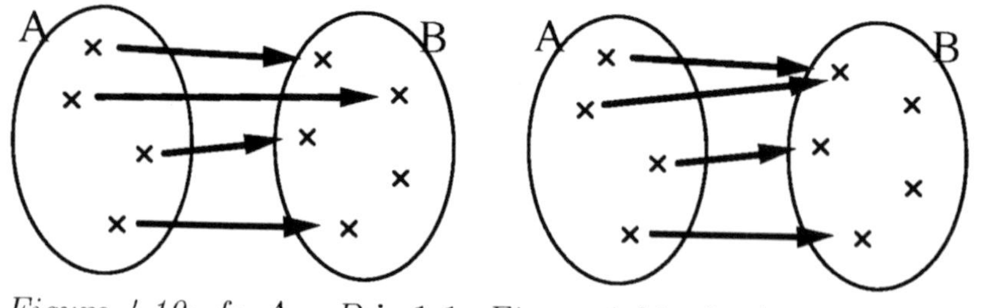


*Figure 4.19:* $f$: $A \to B$ is 1-1 *Figure 4.20:* $f$: $A \to B$ is not 1-1

**Definition.** Let $A$ and $B$ be sets, and let $f: A \to B$. The function $f$ is *onto* (or *surjective*) if for all $b \in B$ there exists $a \in A$ such that $f(a) = b$. An onto function is called a *surjection*.

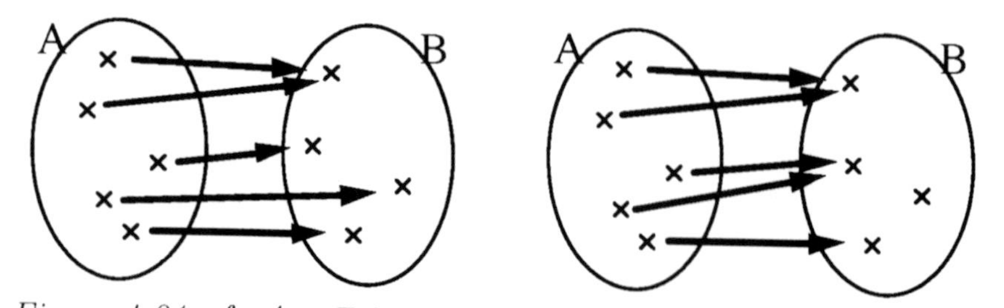


*Figure 4.21:* $f$: $A \to B$ is onto *Figure 4.22:* $f$: $A \to B$ is not onto

**Remark.** When we say that a function $f: A \to B$ is *onto*, the word *onto* is short for *onto B*. As we proved in Exercise 4.3.6, the same function can be written with different targets. Whether or not the function $f: A \to B$ is onto depends on the target $B$. Thus, surjectivity (or ontoness) is not an inherent property of a function, but of the way we choose to write the function.

**Example (4.3)** Let $f: \mathbb{R} \to \mathbb{R}$ be the function $f(x) = x^2$. Let $C = \{x \in \mathbb{R} \mid x \geq 0\}$. Let $g: \mathbb{R} \to C$ be the function $g(x) = x^2$.

Since the domains of $f$ and $g$ are identical and for each $x \in \text{Dom } f$, $f(x) = g(x)$, the functions $f$ and $g$ are the same. Since $\text{Im } f = \text{Im } g = C$ and $C \neq \mathbb{R}$, $f: \mathbb{R} \to \mathbb{R}$ is not onto, but $g: \mathbb{R} \to C$ is onto.

**Remarks.** Notice that for all sets $A$, $B$, for each function $f: A \to B$, the function $f: A \to \text{Im } f$ is onto.

The statement that $f: A \to B$ is onto is equivalent to the statement that $B = \text{Im } f$.

This may seem like a peculiar and not very logical way of talking. Although mathematics is closely tied to logic, mathematical language is at least in part a natural language like any other. Rules of language are determined by custom and usage, rather than by logic alone.

It has turned out historically that "$f: A \to B$ is onto" is not a bad way to say, "$B = \text{Im } f$."

Why not just always choose the target $B$ of a function $f: A \to B$ so that $B = \text{Im } f$? The reason is that it is often rather difficult to find the image (or range) of a function, but very easy to specify a larger target.

For example, consider the function $f: \mathbb{N} \to \{1, 2, 3\}$ defined by: (a) $f(n) = 1$ if $n$ is odd or $n \leq 4$; (b) $f(n) = 2$, if $n$ is even and there exist odd prime numbers $p$, $q \in \mathbb{N}$ such that $p + q = n$; and (c) $f(n) = 3$ if $n$ is even and n $> 4$ and for all odd prime numbers $p$, $q \in \mathbb{N}$, $p + q \neq n$.

Since it is not known whether or not Goldbach's conjecture (that each even integer greater than 4 is the sum of two odd primes, not necessarily distinct) is true, we do not know whether or not $3 \in \text{Im } f$. Thus, we do not know whether or not $f: \mathbb{N} \to \{1, 2, 3\}$ is onto.

The foregoing example is somewhat contrived. But in general determining the image of a function requires effort. Sometimes we wish to name a function without going to the trouble of finding its range. This is why, in general, the target of a function $f: A \to B$ is not necessarily the same as Im $f$.

**Definition.** Let $A$ and $B$ be sets, and let $f: A \to B$. The function $f$ is *bijective* if $f$ is both one-to-one and onto. A bijective function is called a *bijection*, or a *one-to-one correspondence*.

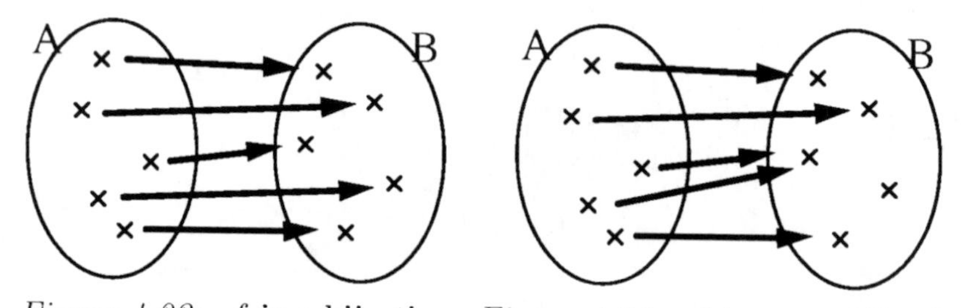


*Figure 4.23:* $f$ is a bijection *Figure 4.24:* $f$ is not a bijection

**Examples (4.4)**

1. Let $f: \mathbb{R} \to \mathbb{R}$ be defined by $f(x) = 7x$. Then $f$ is one-to-one.

   Proof. Let $x$, $y \in \mathbb{R}$. Suppose $f(x) = f(y)$. Then $7x = 7y$. Therefore $x = y$. Thus for all $x$, $y \in \mathbb{R}$, if $f(x) = f(y)$, then $x = y$. That is, $f$ is one-to-one. Q.E.D.

2. Let $f: \mathbb{R} \to \mathbb{R}$ be defined by $f(x) = 7x$. Then $f$ is onto.

   Proof. Let $x \in \mathbb{R}$. Then $\frac{x}{7} \in \mathbb{R}$, and $f(\frac{x}{7}) = 7(\frac{x}{7}) = x$. Thus for all $x \in \mathbb{R}$, there exists $y \in \mathbb{R}$ (namely, $y = \frac{x}{7}$) such that $f(y) = x$. That is, $f$ is onto. Q.E.D.

3. Let $f: \mathbb{Z} \to \mathbb{Z}$ be defined by $f(x) = 7x$. Then $f$ is not onto.

   Proof. The number 2 belongs to $\mathbb{Z}$, and for all $x \in \mathbb{Z}$, $f(x) \neq 2$. Thus there exists $z \in \mathbb{Z}$ (namely, $z = 2$) such that for all $x \in \mathbb{Z}$, $f(x) \neq z$. That is, $f$ is not onto. Q.E.D.

4. Let $f: \mathbb{Z} \to \mathbb{Z}$ be defined by $f(x) = |x|$. Then $f$ is not one-to-one.

   Proof. By definition, $f(3) = |3| = 3$, and $f(-3) = |-3| = 3$. Thus there exist $x, y \in \mathbb{Z}$ (namely, $x = 3$ and $y = -3$) such that $f(x) = f(y)$ but $x \neq y$. That is, $f$ is not one-to-one. Q.E.D.

**Theorem (4.6)** Let $A$ be a set. The identity function $I_A: A \to A$ is a bijection.

**Proof.** Let $x, y \in A$ such that $I_A(x) = I_A(y)$. By definition of $I_A$, $I_A(x) = x$ and $I_A(y) = y$. Hence $x = y$. Thus for all $x, y \in A$, if $I_A(x) = I_A(y)$ then $x = y$. That is, $I_A$ is one-to-one.

Let $a \in A$. Then $I_A(a) = a$. Thus for all $a \in A$ there exists $b \in A$ (namely, $b = a$) such that $I_A(b) = a$. That is, $I_A$ is onto.

Since $I_A$ is one-to-one and onto, $I_A$ is a bijection. Q.E.D.

**Exercises (4.5)** Determine whether each function is (a) one-to-one; (b) onto. Justify your answer. (That is, prove that your answer is correct.)

1. $f: \mathbb{R} \to \mathbb{R}$ defined by $f(x) = 3x - 4$.
2. $f: \mathbb{Z} \to \mathbb{Z}$ defined by $f(x) = 5 - 2x$.
3. $f: \mathbb{Z} \to \mathbb{Z}$ defined by $f(x) = x + 3$.
4. $f: \mathbb{N} \to \mathbb{N}$ defined by $f(x) = x + 3$.
5. $f: \mathbb{R} \to \mathbb{R}$ defined by $f(x) = x^2$.
6. $f: \mathbb{R} \to \mathbb{R}$ defined by $f(x) = x^3$.
7. $f: \mathbb{Z} \times \mathbb{N} \to \mathbb{Q}$ defined by $f(x, y) = \frac{x}{y}$.
8. $f: \mathbb{N} \to \mathbb{Z}$, defined by $f(x) = \begin{cases} \frac{x-1}{2}, & \text{if } x \text{ is odd;} \\ \frac{-x}{2}, & \text{if } x \text{ is even.} \end{cases}$

Prove each theorem, and disprove each false proposition by giving a counterexample.

9. Theorem. Let $A$, $B$, and $C$ be sets. Let $f: A \to B$ and let $g: B \to C$. If $f$ is injective and $g$ is injective, then $g \circ f$ is injective.

10. Theorem. Let $A$, $B$, and $C$ be sets. Let $f: A \to B$ and let $g: B \to C$. If $f$ is surjective and $g$ is surjective, then $g \circ f$ is surjective.

11. Theorem. Let $A$, $B$, and $C$ be sets. Let $f: A \to B$ and let $g: B \to C$. If $f$ and $g$ are both bijections, then $g \circ f$ is a bijection.

12. Theorem. Let $A$, $B$, and $C$ be sets. Let $f: A \to B$ and let $g: B \to C$. If $g \circ f$ is injective, then $f$ is injective.

13. False Proposition. Let $A$, $B$, and $C$ be sets. Let $f: A \to B$ and let $g: B \to C$. If $g \circ f$ is injective, then $g$ is injective.

14. Theorem. Let $A$, $B$, and $C$ be sets. Let $f: A \to B$ and let $g: B \to C$. If $g \circ f$ is surjective, then $g$ is surjective.

15. False Proposition. Let $A$, $B$, and $C$ be sets. Let $f: A \to B$ and let $g: B \to C$. If $g \circ f$ is surjective, then $f$ is surjective.

16. Theorem. Let $X$ and $Y$ be sets, and let $f: X \to Y$. Then $f$ is one-to-one if and only if for all $A, B \subseteq X$, $f(A \cap B) = f(A) \cap f(B)$.

17. Theorem. Let $X$ and $Y$ be sets, and let $f: X \to Y$. Then $f$ is one-to-one if and only if for all $A, B \subseteq X$, $f(A - B) = f(A) - f(B)$.

18. Theorem. Let $X$ and $Y$ be sets, and let $f: X \to Y$. Then $f$ is one-to-one if and only if for all $A \subseteq X$, $f^{-1}(f(A)) = A$.

19. Theorem. Let $X$ and $Y$ be sets, and let $f: X \to Y$. If $f(f^{-1}(Y)) = Y$ then $f$ is onto.

20. Theorem. Let $X$ and $Y$ be sets, and let $f: X \to Y$. If $f$ is onto then for all $A \subseteq Y$, $f(f^{-1}(A)) = A$.

**Theorem (4.7)** Let $A$ and $B$ be sets and let $f\colon A \to B$. The function $f$ is invertible if and only if $f$ is a bijection.

**Proof.** Let $A$ and $B$ be sets and let $f\colon A \to B$. Suppose that $f$ is invertible. Then there exists $f^{-1}\colon B \to A$ such that $f^{-1} \circ f = I_A$ and $f \circ f^{-1} = I_B$. We will show that $f$ is one-to-one and onto.

Let $x$, $y \in A$ such that $f(x) = f(y)$. Since $f(x) = f(y)$, $f^{-1}(f(x)) = f^{-1}(f(y))$. That is, $(f^{-1} \circ f)(x) = (f^{-1} \circ f)(y)$. Hence $I_A(x) = I_A(y)$. Thus $x = y$. Thus, for all $x$, $y \in A$, if $f(x) = f(y)$ then $x = y$. That is, $f$ is one-to-one.

Let $b \in B$. Then $I_B(b) = b$. Since $f \circ f^{-1} = I_B$, $f(f^{-1}(b)) = b$. Let $a = f^{-1}(b)$. Then $f(a) = b$. Thus, for all $b \in B$ there exists $a \in A$ such that $f(a) = b$. That is, $f$ is onto.

Since $f$ is one-to-one and onto, $f$ is a bijection. Thus, if $f$ is invertible then $f$ is a bijection.

Suppose that $f$ is a bijection. We define a function $g\colon B \to A$ as follows. Let $x \in B$. Since $f$ is onto, there exists $y \in A$ such that $f(y) = x$. Since $f$ is one-to-one, there is only one such $y$. Let $g(x) = y$. We claim that $g$ is the inverse function of $f$.

Since Dom $I_A = A$ and Dom $g \circ f = A$, Dom $I_A =$ Dom $g \circ f$. Similarly, Dom $I_B =$ Dom $f \circ g = B$.

Let $a \in A$. By definition, $g(f(a))$ is the unique element $x$ of $A$ such that $f(x) = f(a)$. Hence $x = a$. Thus, for all $a \in A$, $(g \circ f)(a) = a$. Since Dom $I_A =$ Dom $g \circ f$ and for all $a \in A$, $(g \circ f)(a) = I_A(a)$, $g \circ f = I_A$.

Let $b \in B$. Then $g(b)$ is the unique element $y$ of $A$ such that $f(y) = b$. Thus $f(g(b)) = f(y) = b$. Thus, for all $b \in B$, $(f \circ g)(b) = b$. Since Dom $I_B =$ Dom $f \circ g$ and for all $b \in B$, $(f \circ g)(b) = I_B(b)$, $f \circ g = I_B$.

Since $g \circ f = I_A$ and $f \circ g = I_B$, $g = f^{-1}$. Hence $f$ is invertible.

Therefore, $f$ is invertible if and only if $f$ is a bijection. Q.E.D.

Exercises (4.6) Prove each theorem and disprove each false proposition by giving a counterexample.

1. Theorem. Let $f:\mathbb{N} \to \mathbb{N}$, and suppose that for all $x$, $y \in \mathbb{N}$, if $x > y$ then $f(x) > f(y)$. Then $f$ is one-to-one.

2. False Proposition. Let $f:\mathbb{N} \to \mathbb{N}$, and suppose that for all $x$, $y \in \mathbb{N}$, if $f(x) > f(y)$ then $x > y$. Then $f$ is one-to-one.

3. False Proposition. Let $f:\mathbb{N} \to \mathbb{N}$ be one-to-one. Then for all $x$, $y \in \mathbb{N}$, if $x > y$ then $f(x) > f(y)$.

4. Theorem. Let $f:\mathbb{N} \to \mathbb{Z}$ be a bijection. Then there exist $x$, $y \in \mathbb{N}$ such that $x > y$ and $f(x) < f(y)$.

5. False Proposition. Let $f:\mathbb{N} \to \mathbb{N}$ be a function. Then $f$ is one-to-one if and only if $f$ is onto.

# Chapter 5

# Induction, Power Sets, and Cardinality

*Our research firm has devised a simple test whereby we can tell whether or not a given person is a mathematician. Suppose that we have two candidates, A and B. One is a mathematician and one is not, but we do not know which is which.*

*The test requires two days. On the first day, each candidate is put in a modern kitchen, equipped with a table, a stove, and a sink with running water. A teapot, a teacup, and a box of teabags sit on the table. The candidate is asked to make a cup of tea.*

*On the first day, all candidates do the same thing. Each one fills the teapot with tap water, puts it on the stove, and turns the stove on. When the water comes to a boil, each candidate shuts off the stove, puts a teabag in the teacup, and pours the boiling water over the teabag. So much for Day One.*

*On the second day, we invite each candidate separately back to the modern kitchen. But this time, the teapot is on the stove, the stove is on, the water is boiling, and the teabag is already in the teacup. The non-mathematical candidate turns off the stove and pours the boiling water over the teabag.*

*But the mathematician turns off the stove, pours the water out of the teapot into the sink, takes the teabag out of the teacup, and puts the teabag back in its box. This reduces the problem to the previous one, which has already been solved.*

*Old joke*

*Mr Mathews once wrote some sonnets "On Man," and Mr Channing some lines on "A Tin Can," or something of that kind—and if the former gentleman be not the very worst poet that ever existed on the face of the earth, it is only because he is not quite so bad as the latter. To speak algebraically:—Mr M. is* execrable *but Mr C. is* x plus 1*-ecrable.*

*Edgar Allen Poe, reviewing James Russell Lowell*

**Definition.** Let $A$ be a set, and let $\preccurlyeq$ be a partial order on $A$. Let $x \in A$. The element $x$ is a *least element* of $A$ if for all $y \in A$, $x \preccurlyeq y$.

**Definition.** Let $A$ be a set, and let $\preccurlyeq$ be a partial order on $A$. The set $A$ is *well-ordered* by $\preccurlyeq$ if every nonempty subset of $A$ has a least element.

Henceforth we will limit the scope of our discussion to subsets of $\mathbb{R}$ and the partial order $\leq$ .

**Axiom.** (Well-ordering property of the natural numbers) The set of natural numbers $\mathbb{N}$ is well-ordered by $\leq$ . That is, for each subset $S$ of $\mathbb{N}$, if $S$ is nonempty then $S$ has a least element.

**Exercises (5.1)**

1. Theorem. Let $A$ be a subset of $\mathbb{R}$. Then $A$ has at most one least element. [Hint: Let $x, y \in A$ such that $x$ is a least element of $A$ and $y$ is a least element of $A$. Prove that $x = y$.]

2. Theorem. Let $A = \{x \in \mathbb{Q} \mid x > 0\}$. Then $A$ has no least element.

**Theorem (5.1)** (Principle of mathematical induction.) Let $T$ be a subset of $\mathbb{N}$ such that $1 \in T$ and for each $n \in \mathbb{N}$, if $n \in T$ then $n + 1 \in T$. Then $T = \mathbb{N}$.

**Proof.** Let $T$ be a subset of $\mathbb{N}$ such that $1 \in T$ and for each $n \in \mathbb{N}$, if $n \in T$ then $n + 1 \in T$. By way of contradiction, suppose that $T \neq \mathbb{N}$. Since $T \subseteq \mathbb{N}$ and $T \neq \mathbb{N}$, it follows that $\mathbb{N} \not\subseteq T$. Since $\mathbb{N} \not\subseteq T$, there exists $x \in \mathbb{N}$ such that $x \notin T$.

Let $S = \mathbb{N} - T$. Since $x \in S$, the set $S$ is nonempty Since $S$ is a nonempty subset of $\mathbb{N}$, the well-ordering property of the natural numbers implies that $S$ has a least element. Let $m$ be the least element of $S$.

Since $1 \in T$, $1 \notin S$. Hence $m \neq 1$. Since $m \in \mathbb{N}$ and $m \neq 1$, the number $m - 1 \in \mathbb{N}$. Since $m$ is the least element of $S$, it follows that $m - 1 \notin S$. Thus $m - 1 \in T$. By hypothesis, since $m - 1 \in T$, also $(m - 1) + 1 \in T$. Thus $m \in T$. But $m \in S$ and $S = \mathbb{N} - T$. Thus $m \notin T$. Therefore $m \in T$ and $m \notin T$. $\rightarrow\leftarrow$

Our hypothesis has led to a contradiction and is therefore false. Hence, $T = \mathbb{N}$.

Therefore, for each subset $T$ of $\mathbb{N}$, if $1 \in T$ and for each $n \in \mathbb{N}$, $n \in T$ implies $n + 1 \in T$, then $T = N$. Q.E.D.

**Remark.** Since the principle of mathematical induction concerns the set $\mathbb{N}$ of natural numbers, we will use it to prove theorems involving numbers. Our next theorem is about the sum $1 + 2 + 3 \ldots + n$, for each natural number $n$.

Therefore, we will now introduce the summation sign $\Sigma$. The summation sign is a capital sigma. Sigma ($\Sigma$) is the Greek form of the letter S. Sigma is often used to indicate a sum (S is for "sum"), just as delta $(\Delta, \delta)$, the Greek D, is often used to indicate a difference. (Here "difference" means the result of subtracting one number from another.)

**Notation.** Let $f\colon \mathbb{Z} \to \mathbb{R}$ be a function. Let $m$, $n \in \mathbb{Z}$, such that $m < n$. The symbol $\sum_{k=m}^{n} f(k)$ denotes the sum $f(m) + f(m+1) + \ldots + f(n)$.

For each $m \in \mathbb{Z}$, $\sum_{k=m}^{m} f(k) = f(m)$. For each $m$, $n \in \mathbb{N}$, if $m > n$ then $\sum_{k=m}^{n} f(k) = 0$. The symbol $\sum_{k=m}^{n} f(k)$ is pronounced, "the sum, from $k$ equals $m$ to $n$, of $f$ of $k$."

**Remarks.** Some students find the summation sign intimidating at first. Actually it is a rather handy abbreviation for writing down a sum of many terms.

In case you dislike the notation $\sum_{k=m}^{n} f(k)$, feel free to use "$f(m) + f(m+1) + \ldots + f(n)$" in your own work. It is necessary to be able to read all standard mathematical notation. But you need not use notation which you dislike. You will probably come to appreciate the summation sign in time. It really is very convenient for writing long sums.

**Examples (5.1)**

1. $\sum_{k=1}^{5} k = 1 + 2 + 3 + 4 + 5$.

**2.** $\sum_{k=-1}^{5} k^2 = 1 + 0 + 1 + 4 + 9 + 16 + 25 = 56.$

**3.** $\sum_{k=2}^{2} k^2 = 4.$

**4.** $\sum_{k=2}^{1} k^2 = 0.$

**5.** $\sum_{k=1}^{5} 2 = 10.$

**Exercises (5.2)** Evaluate the following sums.

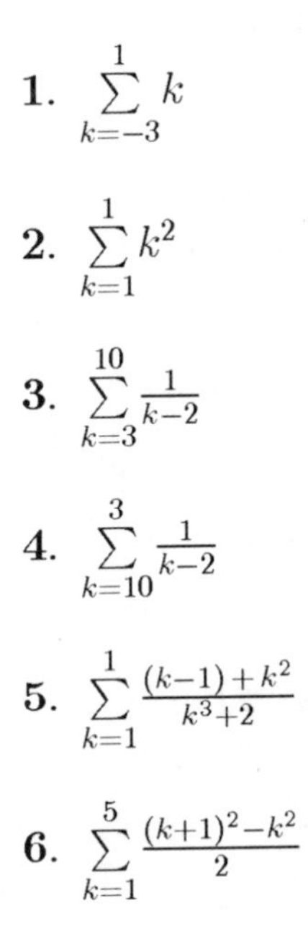

**1.** $\sum_{k=-3}^{1} k$

**2.** $\sum_{k=1}^{1} k^2$

**3.** $\sum_{k=3}^{10} \frac{1}{k-2}$

**4.** $\sum_{k=10}^{3} \frac{1}{k-2}$

**5.** $\sum_{k=1}^{1} \frac{(k-1)+k^2}{k^3+2}$

**6.** $\sum_{k=1}^{5} \frac{(k+1)^2-k^2}{2}$

Now we are ready to prove a theorem using the principle of mathematical induction.

**Theorem (5.2)** For each $n \in \mathbb{N}$, $\sum_{k=1}^{n} k = \frac{n(n+1)}{2}$.

**Proof.** The proof is by induction on $n$.

Let $T = \{n \in \mathbb{N} \mid \sum_{k=1}^{n} k = \frac{n(n+1)}{2}\}$.

Base Step. Since $\sum_{k=1}^{1} = 1$ and $\frac{1(1+1)}{2} = 1$, $\sum_{k=1}^{1} = \frac{1(1+1)}{2}$. Hence $1 \in T$.

Induction Step. Let $n \in \mathbb{N}$ and suppose that $n \in T$. We will show that $n+1 \in T$. The sum $\sum_{k=1}^{n+1} k = \sum_{k=1}^{n} k + (n+1)$. Since $n \in T$, $\sum_{k=1}^{n} k = \frac{n(n+1)}{2}$. Substituting, we get $\sum_{k=1}^{n+1} k = \frac{n(n+1)}{2} + (n+1)$. Since $\frac{n(n+1)}{2} + (n+1) = \frac{n(n+1)}{2} + \frac{2(n+1)}{2} = \frac{(n+2)(n+1)}{2} = \frac{(n+1)((n+1)+1)}{2}$, we conclude that $\sum_{k=1}^{n+1} k = \frac{(n+1)((n+1)+1)}{2}$. Hence $n+1 \in T$. Thus, for each $n \in \mathbb{N}$, if $n \in T$ then $n+1 \in T$.

Since $1 \in T$ and since, for each $n \in \mathbb{N}$, if $n \in T$ then $n+1 \in T$, it follows by the principle of mathematical induction that $T = \mathbb{N}$.

Therefore, by the principle of mathematical induction, for each $n \in \mathbb{N}$, $\sum_{k=1}^{n} k = \frac{n(n+1)}{2}$. Q.E.D.

**Remarks.** First we show that $1 \in T$. This is called the *base step*, or the *basis* for the induction.

Next we show that for each $n \in \mathbb{N}$, if $n \in T$ then $n+1 \in T$. This is called the *induction step*. We write: "Let $n \in \mathbb{N}$ and suppose that $n \in T$." The hypothesis that $n \in T$ (that is, that $\sum_{k=1}^{n} k = \frac{n(n+1)}{2}$) is called the *induction hypothesis*.

Notice that in the induction step, we do not proceed as follows. "Let $n \in \mathbb{N}$ and suppose that $\sum_{k=1}^{n} k = \frac{n(n+1)}{2}$. Set $\sum_{k=1}^{n+1} k = \frac{(n+1)(n+2)}{2} = \frac{n^2+3n+2}{2} = \frac{n^2+n}{2} + \frac{2n+2}{2} = \sum_{k=1}^{n} k + n + 1$." Here we are beginning with what we intend to show (namely, that $\sum_{k=1}^{n+1} k = \frac{(n+1)((n+1)+1)}{2}$) and working backward until we reach the induction hypothesis (namely, that $\sum_{k=1}^{n} k = \frac{n(n+1)}{2}$). This is tantamount to showing that for each $n \in \mathbb{N}$, if $n+1 \in T$, then $n \in T$. But what we actually need to show is that for each $n \in \mathbb{N}$, if $n \in T$ then $n+1 \in T$.

Writing induction proofs backwards is a very common mistake among beginners. This is probably because working backwards often helps the mathematician to see whether the theorem is true and, if so, how to prove it. When you are writing induction proofs, feel free to work backwards on your scrap paper, to help yourself figure out your strategy. But in your finished proof the induction step must begin with what you are assuming (the induction hypothesis that $n \in T$) and end with what you are proving (that $n+1 \in T$).

## Exercises (5.3)

1. Theorem. For all $n \in \mathbb{N}$, $\sum_{k=1}^{n} k^2 = \frac{n(n+1)(2n+1)}{6}$.

2. Theorem. For all $n \in \mathbb{N}$, $\sum_{k=1}^{n} (2k-1) = n^2$.

3. Theorem. For all $n \in \mathbb{N}$, $\sum_{k=1}^{n} \frac{1}{k(k+1)} = \frac{n}{n+1}$.

4. Theorem. For all $n \in \mathbb{N}$, $\sum_{k=1}^{n} k(k+1) = \frac{n(n+1)(n+2)}{3}$.

5. Theorem. For all $n \in \mathbb{N}$, $\sum_{k=1}^{n} k^3 = \frac{n^2(n+1)^2}{4}$.

6. Theorem. For all $n \in \mathbb{N}$, $2^n \geq n+1$.

7. Theorem. For all $n \in \mathbb{N}$, $3^n \geq 2^n + 1$.

The technique of mathematical induction has become so familiar to mathematicians that, most of the time, we do not explicitly introduce or allude to the set $T$ when writing induction proofs. (We may introduce the set $T$ if we wish.) Thus, the proof of Theorem 5.2 would usually be written more or less as follows:

**Theorem (5.3)** For all $n \in \mathbb{N}$, $\sum_{k=1}^{n} k = \frac{n(n+1)}{2}$.

**Proof.** The proof is by induction on $n$.

Let $n = 1$. Then $\sum_{k=1}^{n} k = \sum_{k=1}^{1} k = 1$. Since $n = 1$, $\frac{n(n+1)}{2} = \frac{1(1+1)}{2} = 1$. Since $\sum_{k=1}^{1} k = 1$ and $\frac{1(1+1)}{2} = 1$, $\sum_{k=1}^{1} k = \frac{1(1+1)}{2}$.

Let $n \in \mathbb{N}$ and suppose that $\sum_{k=1}^{n} k = \frac{n(n+1)}{2}$. Since $\sum_{k=1}^{n+1} k = \sum_{k=1}^{n} k + (n+1)$ and $\sum_{k=1}^{n} k = \frac{n(n+1)}{2}$, we find that $\sum_{k=1}^{n+1} k = \frac{n(n+1)}{2} + (n+1)$. Since $\frac{n(n+1)}{2} + (n+1) = \frac{n(n+1)}{2} + \frac{2(n+1)}{2} = \frac{(n+2)(n+1)}{2} = \frac{(n+1)((n+1)+1)}{2}$, it follows that $\sum_{k=1}^{n+1} k = \frac{(n+1)((n+1)+1)}{2}$.

Therefore, by the principle of mathematical induction, for all $n \in \mathbb{N}$, $\sum_{k=1}^{n} k = \frac{n(n+1)}{2}$. Q.E.D.

**Exercises (5.4)**

1. Theorem. For each $n \in \mathbb{N}$, $\sum_{k=1}^{2n} (-1)^k k = n$.

2. Theorem. For each $n \in \mathbb{N}$, for each $x \in \mathbb{R}$, if $x \neq 1$ then $\sum_{k=0}^{n} x^k = \frac{x^{n+1}-1}{x-1}$.

3. Theorem. For each $n \in \mathbb{N}$, for each $x \in \mathbb{R}$, if $x \neq -1$ then $\sum_{k=0}^{2n} (-1)^k x^k = \frac{x^{2n+1}+1}{x+1}$.

4. Theorem. For each $n \in \mathbb{N}$, for each $x \in \mathbb{R}$, if $x \neq -1$ then $\sum_{k=0}^{2n-1} (-1)^{k+1}x^k = \frac{x^{2n}+1}{x+1}$.

5. Theorem. For each $n \in \mathbb{N}$, $\sum_{k=1}^{n} (-1)^k k^2 = (-1)^n \left(\frac{n(n+1)}{2}\right)$.

6. Theorem. Let $T \subseteq \mathbb{Z}$ and let $x \in T$. Suppose that for each $k \in T$, $k + 1 \in T$. Then $\{m \in \mathbb{Z} \mid m \geq x\} \subseteq T$.

7. Theorem. For each $n \in \mathbb{N}$ such that $n \geq 3$, $2^n > 2n$. [Hint: Use Exercise 6. That is, use induction with the case $n = 3$ as the base step.]

8. Theorem. For each $n \in \mathbb{N}$ such that $n \geq 2$, $3^n > 2^{n+1}$.

9. Theorem. Let $A$ be a subset of $\mathbb{Z}$. Suppose that there exists $x \in A$ such that, for each $T \subseteq A$, if $x \in T$ and if for each $y \in T$, $y + 1 \in T$, then $T = A$. Then $x$ is the least element of $A$.

**Definition.** Let $A$, $B$ be sets, let $f: A \to B$, and let $C \subseteq A$. Let $g: C \to B$ be defined by $g(x) = f(x)$. Then $g$ is the *restriction of $f$ to $C$*, or *$f$ restricted to $C$*. The restriction of $f$ to $C$ is denoted by $f|C$.

**Notation.** For each $k \in \mathbb{N}$, we will use the symbol $\mathbb{N}_k$ to denote the set $\{m \in \mathbb{N} \mid m \leq k\}$. Oddly enough, there is no widely used standard notation for this set. Since, unlike $\mathbb{N}$, $\mathbb{Z}$, and $\mathbb{R}$, the symbol $\mathbb{N}_k$ is not standard notation, be sure to define it when you use it outside this course.

The following exercises will help us prove Theorem 5.4.

**Exercises (5.5)**

1. Theorem. Let $A$, $B$ be sets, let $C \subseteq A$, and let $f: A \to B$. If $f$ is one-to-one, then $f|C$ is one-to-one.

2. Theorem. Let $A$, $B$ be sets, let $C \subseteq A$, and let $f: A \to B$. If $f|C$ is onto, then $f$ is onto.

3. False Proposition. Let $A$, $B$ be sets, let $C \subseteq A$, and let $f: A \to B$. If $f$ is onto, then $f|C$ is onto.

4. Theorem. Let $A$ be a set, let $x \in A$, and let $f: A \to A$. If $f$ is one-to-one and $f(x) = x$, then $f(A - \{x\}) \subseteq A - \{x\}$.

5. Theorem. Let $A$ be a set, let $x \in A$, and let $f: A \to A$. If $f(x) = x$ and $f|(A - \{x\})$ is a bijection on $A - \{x\}$, then $f$ is a bijection on $A$.

6. Theorem. Let $A$ be a set, let $x \in A$, and let $f: A \to A$. Let $h: A \to A$ be defined by $h(a) = \begin{cases} x, \text{ if } a = f(x); \\ f(x) \text{ if } a = x; \\ a \text{ otherwise.} \end{cases}$ Then $h$ is a bijection and $h^{-1} = h$ and $(h \circ f)(x) = x$.

7. Theorem. Let $A, B \subseteq \mathbb{N}$, and let $f: A \to B$ be onto. Then there exists $g: B \to A$ defined by $g(x) =$ the least element of $f^{-1}(\{x\})$. Then $g$ is one-to-one, and $f \circ g = I_B$.

8. Theorem. Let $A$, $B$ be sets, and let $f: A \to B$. If there exists $g: B \to A$ such that $g$ is onto and $f \circ g = I_B$, then $f$ is one-to-one.

**Theorem (5.4)** For each $n \in \mathbb{N}$, for each $f: \mathbb{N}_n \to \mathbb{N}_n$, $f$ is one-to-one if and only if $f$ is onto.

**Proof.** The proof is by induction on $n$.

Let $f: \mathbb{N}_1 \to \mathbb{N}_1$ be a function. Then Dom $f = \{1\}$, and $f(1) = 1$. Since Dom $f =$ Dom $I_{\mathbb{N}_1}$ and for each $x \in$ Dom $f$, $f(x) = I_{\mathbb{N}_1}(x)$, $f = I_{\mathbb{N}_1}$. Thus $f$ is both one-to-one and onto. Therefore, for each $f: \mathbb{N}_1 \to \mathbb{N}_1$, $f$ is one-to-one if and only if $f$ is onto.

Let $n \in \mathbb{N}$, and suppose that for each $g: \mathbb{N}_n \to \mathbb{N}_n$, $g$ is one-to-one if and only if $g$ is onto. Let $f: \mathbb{N}_{n+1} \to \mathbb{N}_{n+1}$. Suppose that $f$ is one-to-one. Either $f(n+1) = n+1$, or not.

Case 1. Suppose that $f(n+1) = n+1$. Since $f$ is one-to-one and $f(n+1) = n+1$, by Exercise 4 $f(\mathbb{N}_n) \subseteq \mathbb{N}_n$. Thus $f|\mathbb{N}_n$ is a function from $\mathbb{N}_n$ to $\mathbb{N}_n$. Since $f$ is one-to-one, by Exercise 1 $f|\mathbb{N}_n$ is one-to-one. Since $f|\mathbb{N}_n$ is a one-to-one function from $\mathbb{N}_n$ to $\mathbb{N}_n$, by the induction hypothesis $f|\mathbb{N}_n$ is onto. Thus $f|\mathbb{N}_n$ is a bijection on $\mathbb{N}_n$. Since $f|\mathbb{N}_n$ is a bijection on $\mathbb{N}$ and $f(n+1) = n+1$, by Exercise 5 $f$ is a bijection on $\mathbb{N}_{n+1}$. Since $f$ is a bijection, $f$ is onto.

Case 2. Suppose that $f(n+1) \neq n+1$. By Exercise 6, there exists a bijection $h: \mathbb{N}_{n+1} \to \mathbb{N}_{n+1}$ such that $(h \circ f)(n+1) = n+1$ and $h = h^{-1}$. Let $g = h \circ f$. Since $f$ is one-to-one and $h$ is one-to-one, $g$ is one-to-one. Thus $g: \mathbb{N}_{n+1} \to \mathbb{N}_{n+1}$ is one-to-one and $g(n+1) = n+1$. Therefore, by Case 1, $g$ is onto. Since $g$ is onto and $h$ is onto, $h \circ g$ is onto. Since $h = h^{-1}$, $h \circ g = h \circ (h \circ f) = (h \circ h) \circ f = I_{\mathbb{N}_{n+1}} \circ f = f$. Since $h \circ g$ is onto and $h \circ g = f$, $f$ is onto.

Therefore, if $f: \mathbb{N}_{n+1} \to \mathbb{N}_{n+1}$ is one-to-one, then $f$ is onto.

It remains to show that if $f$ is onto, then $f$ is one-to-one. Suppose that $f$ is onto. Then by Exercise 7 there exists a one-to-one function $g: \mathbb{N}_{n+1} \to \mathbb{N}_{n+1}$ such that $f \circ g = I_{\mathbb{N}_{n+1}}$. Since every one-to-one function from $\mathbb{N}_{n+1}$ to $\mathbb{N}_{n+1}$ is onto, $g$ is onto. Since $g$ is onto and $f \circ g = I_{\mathbb{N}_{n+1}}$, by Exercise 8 $f$ is one-to-one.

Therefore, if $f: \mathbb{N}_{n+1} \to \mathbb{N}_{n+1}$ is onto, then $f$ is one-to-one.

Hence, for each $f: \mathbb{N}_{n+1} \to \mathbb{N}_{n+1}$, $f$ is one-to-one if and only if $f$ is onto.

Therefore, by the principle of mathematical induction, for all $n \in \mathbb{N}$, for each $f: \mathbb{N}_n \to \mathbb{N}_n$, $f$ is one-to-one if and only if $f$ is onto. Q.E.D.

**Theorem (5.5)** Let $n$, $k \in \mathbb{N}$ and let $f: \mathbb{N}_n \to \mathbb{N}_k$. If $n > k$ then $f$ is not one-to-one.

**Proof.** By way of contradiction, suppose that there exist $n, k \in \mathbb{N}$ and $f: \mathbb{N}_n \to \mathbb{N}_k$ such that $n > k$ and $f$ is one-to-one. Since $f$ is one-to-one, $f|\mathbb{N}_k$ is one-to-one. Since $f|\mathbb{N}_k$ is a one-to-one function from $\mathbb{N}_k$ to $\mathbb{N}_k$, $f|\mathbb{N}_k$ is onto. Since $f(n) \in \mathbb{N}_k$ and $f|\mathbb{N}_k$ is onto, there exists $x \in \mathbb{N}_k$ such that $f(x) = f(n)$. Since $x \in \mathbb{N}_k$ and $n > k$, $x \neq n$. Since $x \neq n$ and $f(x)$

$= f(n)$, $f$ is not one-to-one. But by hypothesis $f$ is one-to-one. Thus $f$ is one-to-one and $f$ is not one-to-one. $\rightarrow\leftarrow$

Our hypothesis has led to a contradiction and is therefore false. Therefore, for all $n, k \in \mathbb{N}$, for each $f: \mathbb{N}_n \rightarrow \mathbb{N}_k$, if $n > k$ then $f$ is not one-to-one. Q.E.D.

**Theorem (5.6)** Let $n, k \in \mathbb{N}$ and let $f: \mathbb{N}_n \rightarrow \mathbb{N}_k$. If $n < k$ then $f$ is not onto.

**Proof.** Exercise.

**Theorem (5.7)** Let $n, k \in \mathbb{N}$. There exists a one-to-one function $f: \mathbb{N}_n \rightarrow \mathbb{N}_k$ if and only if $n \leq k$. There exists an onto function $g: \mathbb{N}_n \rightarrow \mathbb{N}_k$ if and only if $n \geq k$.

**Proof.** Exercise.

**Corollary.** Let $n, k \in \mathbb{N}$. There exists a bijection $f: \mathbb{N}_n \rightarrow \mathbb{N}_k$ if and only if $n = k$.

**Proof.** Exercise.

**Exercises (5.6)**

1. Prove Theorem 5.6.

2. Prove Theorem 5.7.

3. Prove the Corollary to Theorem 5.7.

4. Theorem. Let $A$ be a set and let $n, k \in \mathbb{N}$. If there exist bijections $f: \mathbb{N}_n \rightarrow A$ and $g: \mathbb{N}_k \rightarrow A$ then $n = k$.

**The pigeonhole principle.** We have just gone to some trouble to prove by induction that for all $n \in \mathbb{N}$, for each $f: \mathbb{N}_n \to \mathbb{N}_n$, $f$ is one-to-one if and only if $f$ is onto. From this result we have deduced that for all $n$, $k \in \mathbb{N}$, there exists a one-to-one function $f: \mathbb{N}_n \to \mathbb{N}_k$ if and only if $n \leq k$, and there exists an onto function $g: \mathbb{N}_n \to \mathbb{N}_k$ if and only if $n \geq k$.

If we just draw some pictures, these results seem very obvious. (But let's remember that "obvious" is not well defined.)

**(a)** Let $n$, $k \in \mathbb{N}$. Suppose that $k < n$. Let $f: \mathbb{N}_n \to \mathbb{N}_k$.

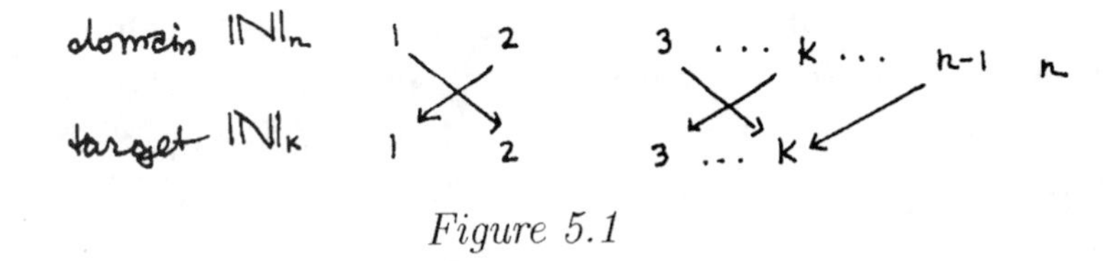


*Figure 5.1*

To represent $f$, we draw an arrow from each number on the top row to a number on the bottom row. Since $k < n$, there are more numbers on top than on the bottom. Hence, two arrows will have to end in the same number. That is, $f$ is not one-to-one.

**(b)** Let $n$, $k \in \mathbb{N}$. Suppose that $k > n$. Let $f: \mathbb{N}_n \to \mathbb{N}_k$.

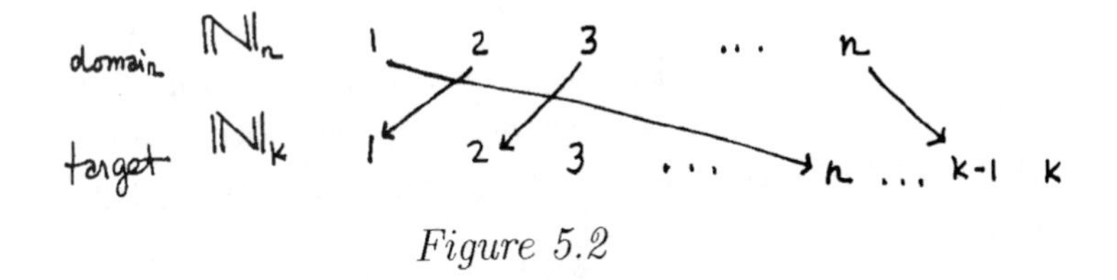


*Figure 5.2*

Since $k > n$, there are not enough numbers in the top row to send arrows to all the numbers on the bottom. So at least one number on the bottom will be left out of Im $f$. That is, $f$ is not onto.

**(c)** Let $n \in \mathbb{N}$ and let $f: \mathbb{N}_n \to \mathbb{N}_n$.

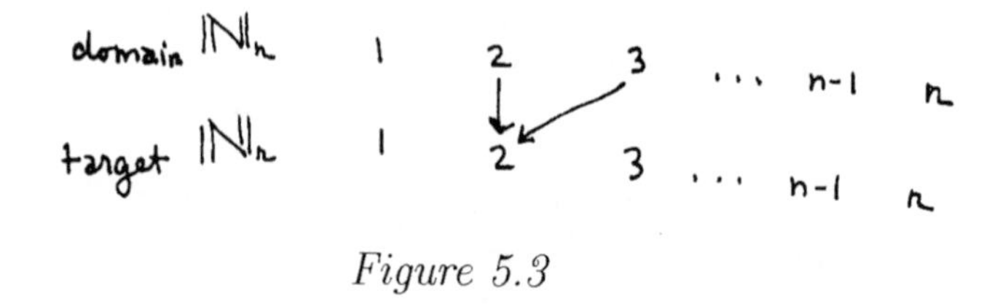


*Figure 5.3*

Suppose $f$ is not one-to-one. Then two numbers from the top have arrows to the same number on the bottom. There won't be enough numbers left on top for all the remaining numbers on the bottom. At least one number on the bottom will be left out of Im $f$. That is, $f$ is not onto.

**(d)** Let $n \in \mathbb{N}$ and let $f: \mathbb{N}_n \to \mathbb{N}_n$.

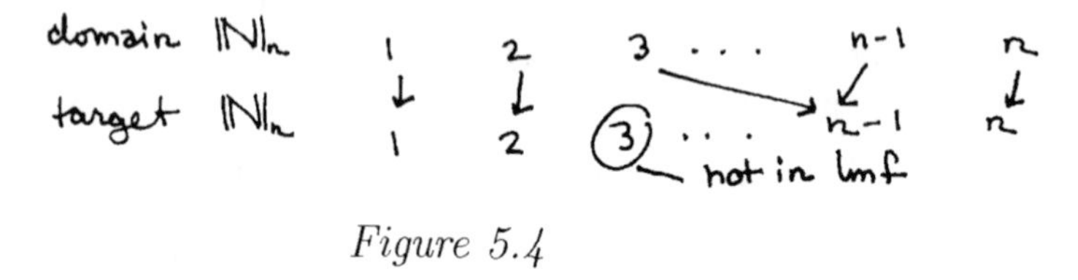


*Figure 5.4*

Suppose $f$ is not onto. Then at least one number on the bottom gets left out of Im $f$. Now there are fewer available numbers on the bottom than on the top. So two arrows from the top must end in the same number on the bottom.

Thus, the results of Theorems 5.4, 5.5, 5.6, and 5.7 are often regarded as "common sense," and are collectively known as "the pigeonhole principle." If there are more pigeons than pigeonholes, and every pigeon gets a pigeonhole, then some pigeons will have to share.

Why, in this case, bother to prove Theorem 5.4 by induction? First of all, the induction proof is cool. The induction proof is a bit more satisfying than what mathematicians call "handwaving." "Handwaving" means, more or less, appeal to common sense. We mathematicians prefer not to have to appeal to common sense, as there is no universal agreement on exactly what common sense tells us. Secondly, the induction proof uses techniques (e.g., making up functions to compose with existing functions, considering restrictions of functions) that will come in handy in future proofs.

**Counting.** We have reached the point in this book where we wish to discuss counting. What do we mean by counting? What do we mean by the number of elements in a set?

Let $A$ be a set, and let $n \in \mathbb{N}$. Suppose that $A$ has exactly $n$ elements. What does this mean? It means, more or less, that we can number the elements of $A$ as $a_1, a_2, \ldots, a_n$. It means that we can write $A$ as a finite sequence of distinct terms $\{a_k\}_{k=1}^{n}$. But this is really the same thing as defining a function $f: \mathbb{N}_n \to A$ by $f(k) = a_k$. Since Im $f = A$, $f$ is onto. Since for each $m, k \in \mathbb{N}_n$, if $m \neq k$ then $a_m \neq a_k$, $f$ is one-to-one. Thus, $f$ is a bijection from $\mathbb{N}_n$ to $A$.

In other words, for each $n \in \mathbb{N}$, to count the elements of a set $A$ with exactly $n$ elements is to define a bijection $f: \mathbb{N}_n \to A$ by setting $f(k)$ equal to the element of $A$ that comes $k$th in the counting.

In Exercise 4, the reader proved that for each set $A$, if there exists $n \in \mathbb{N}$ such that we can define a bijection $f: \mathbb{N}_n \to A$, then there is only one such $n$. In other words, if we can count the elements of $A$ in this way, we always get the same answer. This is the justification for our next definition.

**Definition.** Let $A$ be a set. If there exist $n \in \mathbb{N}$ and $f: \mathbb{N}_n \to A$ such that $f$ is a bijection, then $n$ is the *number of elements* in $A$, or the *cardinality* of $A$. The statement that $A$ has exactly $n$ elements (or that the cardinality of $A$ is $n$) is denoted by $|A| = n$. The cardinality of the empty set is zero.

**Notation.** Since $|\varnothing| = 0$, in discussing cardinalities of sets we will often refer to the set of nonnegative integers $\mathbb{N} \cup \{0\}$. This set goes by various names in various books. We will refer to the set $\mathbb{N} \cup \{0\}$ by the name most familiar to us and least ambiguous. This name is $\omega_0$ (pronounced "omega nought," and not, for example, "woe"). If the symbol $\omega_0$ and the spoken words "omega nought" seem strange to you, just use them for a while, and you'll get used to them.

Thus, from now on, we will use $\omega_0$ (omega nought) to denote the set of nonnegative integers $\{0, 1, 2, 3 \ldots\}$.

**Definition.** Let $A$ be a set. Then $A$ is *finite* if there exists $k \in \omega_0$ such that $|A| = k$. If $A$ is not finite, then $A$ is *infinite.*

**Exercises (5.7)**

1. Theorem. Let $A$ and $B$ be finite sets. Then $|A| = |B|$ if and only if there exists a bijection $f: A \to B$.

2. Theorem. Let $A$ and $B$ be sets. If $A$ is finite and there exists a bijection $f: A \to B$, then B is finite.

3. Theorem. Let $\mathscr{S}$ be a collection of finite sets. Let $\sim$ be the relation on $\mathscr{S}$ such that $X \sim Y$ if and only if $|X| = |Y|$. Then $\sim$ is an equivalence relation on $\mathscr{S}$.

**Extending the concept of cardinality.** Exercises 5.7 suggest a way of extending the concept of cardinality to apply to all sets, not just to finite sets. Clearly the equivalence relation $\sim$ of Exercise 3 can be extended to sets that are not finite.

**Theorem (5.8)** Let $\mathscr{S}$ be a collection of sets, and let $\sim$ be the relation on $\mathscr{S}$ such that $X \sim Y$ if and only if there exists a bijection $f: X \to Y$. Then $\sim$ is an equivalence relation on $\mathscr{S}$.

**Proof.** Exercise.

**Notation.** Let $\mathscr{S}$ be a collection of sets, and let $\sim$ be the relation of Theorem 5.8. For all finite sets $A$, $B \in \mathscr{S}$, $A \sim B$ if and only if $|A| = |B|$. Thus, it is natural to extend this notation to all sets, finite or not, in $\mathscr{S}$. Thus for all $X$, $Y \in \mathscr{S}$, we write $|X| = |Y|$ to denote the relation $X \sim Y$ of Theorem 5.8. Since, for all finite sets $A$, $B$, the notation $|A| = |B|$ is spoken as, "$A$ and $B$ are equal in cardinality," we use the same words for sets in general. Thus we define the relation "are equal in cardinality" without first defining the noun "cardinality" for infinite sets.

**Definition.** Let $X$ and $Y$ be sets. Then $X$ and $Y$ are *equal in cardinality*, denoted by $|X| = |Y|$, if there exists a bijection $f: X \to Y$.

**Example (5.2)** Let $X$ be the set $\{k \in \mathbb{N} \mid \text{there exists } m \in \mathbb{N} \text{ such that } k = 2m\}$. Then $X$ is the set $\{2, 4, 6, \ \dots\}$. Let $f: \mathbb{N} \to X$ be defined by $f(n) = 2n$. Then $f$ is a bijection. Therefore, $|X| = |\mathbb{N}|$.

**Remarks.** In the example above, $X \subseteq \mathbb{N}$ and $X \neq \mathbb{N}$ and $|X| = |\mathbb{N}|$. Notice that by Theorem 5.4, this could not happen if $\mathbb{N}$ were a finite set.

Mathematicians do define transfinite cardinal numbers which are the cardinalities of infinite sets. If you take a course in advanced set theory or point-set topology (see Afterword D), you will learn about transfinite numbers. But they fall outside the scope of this book. Afterword A contains a very brief discussion of transfinite numbers.

**Theorem (5.9)** Let $A$, $B$ be finite sets. If $A \cap B = \varnothing$ then $|A \cup B| = |A| + |B|$.

**Proof.** Let $A$, $B$ be finite sets such that $A \cap B = \varnothing$. Either both sets are nonempty or $A$ is empty or $B$ is empty.

Case 1. Suppose that $A \neq \varnothing$ and $B \neq \varnothing$. Then there exist $n$, $k \in \mathbb{N}$ such that $|A| = n$ and $|B| = k$. Since $|A| = n$ and $|B| = k$, there exist bijections $f: A \to \mathbb{N}_n$ and $g: B \to \mathbb{N}_k$. We define a function

$$h: A \cup B \to \mathbb{N}_{n+k} \text{ by } h(x) = \begin{cases} f(x), \text{ if } x \in A; \\ n + g(x) \text{ otherwise.} \end{cases}$$

We leave it to the reader as an exercise to prove that $h$ is a bijection. Since $h$ is a bijection from $A \cup B$ to $\mathbb{N}_{n+k}$, $|A \cup B| = n + k = |A| + |B|$.

Case 2. Suppose that $A = \varnothing$. Since $A = \varnothing$, $A \cup B = B$. Thus $|A \cup B| = |B| = 0 + |B| = |A| + |B|$.

Case 3. Suppose that $B = \varnothing$. Since $B = \varnothing$, $A \cup B = A$. Thus $|A \cup B| = |A| = |A| + 0 = |A| + |B|$.

Therefore, for all finite sets $A$, $B$, if $A \cap B = \varnothing$ then $|A \cup B| = |A| + |B|$. Q.E.D.

**Definition.** Let $A$ be a set. The *power set* of $A$, denoted by $\mathcal{P}(A)$, is the set of all subsets of $A$.

**Examples (5.3)**

1. Let $A = \{1, 2, 3\}$. The power set of $A$ is the set $\{\{1, 2, 3\}, \{1, 2\}, \{1, 3\}, \{2, 3\}, \{1\}, \{2\}, \{3\}, \varnothing\}$.
2. The power set of the empty set $\varnothing$ is the set $\{\varnothing\}$. The set $\{\varnothing\}$ contains one element (namely $\varnothing$), and hence is not the same as the empty set $\varnothing$, which has no elements.
3. Power sets themselves have power sets. Let $A = \{1\}$. Then $\mathcal{P}(A) = \{\{1\}, \varnothing\}$, and $\mathcal{P}(\mathcal{P}(A)) = \{\{\{1\}, \varnothing\}, \{\{1\}\}, \{\varnothing\}, \varnothing\}$.
4. The set $\mathcal{P}(\mathbb{N})$ is the power set of $\mathbb{N}$. Hence $\mathbb{N} \in \mathcal{P}(\mathbb{N})$, $\varnothing \in \mathcal{P}(\mathbb{N})$, $\{1, 2, 3\} \in \mathcal{P}(\mathbb{N})$, $\{n \in \mathbb{N} \mid \text{there exists } m \in \mathbb{N} \text{ such that } n = 3m\} \in \mathcal{P}(\mathbb{N})$, and so on.

**Exercises (5.8)**

1. Theorem. Let $A$, $B$ be disjoint nonempty finite sets. Let $n$, $k \in \mathbb{N}$, and let $f: A \to \mathbb{N}_n$ and $g: B \to \mathbb{N}_k$ be bijections, and let $h: A \cup B \to \mathbb{N}_{n+k}$ be defined by $h(x) = \begin{cases} f(x), & \text{if } x \in A; \\ n + g(x) & \text{otherwise.} \end{cases}$

   Then $h$ is a bijection.

2. Let $A = \{5, 7\}$. Find $\mathcal{P}(A)$ and $\mathcal{P}(\mathcal{P}(A))$.

3. Find $\mathcal{P}(\mathcal{P}(\varnothing))$ and $\mathcal{P}(\mathcal{P}(\mathcal{P}(\varnothing)))$.

4. Theorem. Let $A$ and $B$ be sets. If $B \subseteq A$ then $\mathcal{P}(B) \subseteq \mathcal{P}(A)$.

5. Theorem. Let $A$ and $B$ be sets. Then $\mathcal{P}(A) \cap \mathcal{P}(B) = \mathcal{P}(A \cap B)$.

6. False Proposition. For all sets $A$ and $B$, $\mathcal{P}(A \cup B) = \mathcal{P}(A) \cup \mathcal{P}(B)$.

7. False Proposition. For all sets $A$ and $B$, $\mathcal{P}(A \times B) = \mathcal{P}(A) \times \mathcal{P}(B)$.

8. False Proposition. For all sets $A$ and $B$, $\mathcal{P}(A - B) = \mathcal{P}(A) - \mathcal{P}(B)$.

9. Theorem. Let $f: \mathbb{N} \to \mathcal{P}(\mathbb{N})$ be defined by $f(n) = \{n\}$. Then $f$ is one-to-one but not onto.

10. Theorem. Let $f: \mathcal{P}(\mathbb{N}) \to \mathbb{N}$ be defined as follows. If $S = \varnothing$, then $f(S) = 37$. If $S \neq \varnothing$, then $f(S)$ is the least element of $S$. Then $f$ is onto but not one-to-one.

11. Theorem. Let $f: \mathcal{P}(\mathbb{N}) \to \mathcal{P}(\mathbb{N})$ be defined by $f(S) = \mathbb{N} - S$. Then $f$ is a bijection.

12. Give an example of a function $f$: $\mathcal{P}(\mathbb{N}) \to \mathcal{P}(\mathbb{N})$ such that $f$ is one-to-one but not onto.

13. Give an example of a function $f$: $\mathcal{P}(\mathbb{N}) \to \mathcal{P}(\mathbb{N})$ such that $f$ is onto but not one-to-one.

14. Theorem. Let $A$ and $B$ be nonempty sets and let $h: A \to B$ be a bijection. Let $H$: $\mathcal{P}(A) \to \mathcal{P}(B)$ be defined by $H(S) = h(S)$. If $h$ is a bijection then $H$ is a bijection.

15. Let $f: \mathbb{N} \to \mathcal{P}(\mathbb{N})$ be defined by $f(n) = \begin{cases} \mathbb{N}_n & \text{if } n \text{ is odd;} \\ \mathbb{N} - \mathbb{N}_n & \text{if } n \text{ is even.} \end{cases}$

    Let $G$ be the set $\{n \in \mathbb{N} \mid n \notin f(n)\}$. Find $G$. Is $G \in \operatorname{Im} f$?

**Theorem (5.10)** For each set $A$, for each function $f: A \to \mathcal{P}(A)$, the function $f$ is not onto.

**Proof.** By way of contradiction, suppose that there exist a set $A$ and a function $f: A \to \mathcal{P}(A)$ such that $f$ is onto.

Let $G = \{x \in A \mid x \notin f(x)\}$. Since $G \in \mathcal{P}(A)$ and $f$ is onto, there exists $y \in A$ such that $f(y) = G$. Either $y \in G$ or $y \notin G$.

Case 1. Suppose that $y \in G$. Then, by the definition of $G$, $y \notin f(y)$. Since $f(y) = G$ and $y \notin f(y)$, it follows that $y \notin G$. Thus $y \in G$ and $y \notin G$. $\to\leftarrow$

Case 2. Suppose that $y \notin G$. Then, by the definition of $G$, $y \in f(y)$. Since $f(y) = G$ and $y \in f(y)$, it follows that $y \in G$. Thus $y \notin G$ and $y \in G$. $\to\leftarrow$

The hypothesis that $f$ is onto has led to a contradiction. Therefore, for each set $A$, for each function $f: A \to \mathcal{P}(A)$, the function $f$ fails to be onto. Q.E.D.

**Remarks.** Let $n \in \mathbb{N}$ and let $A$ be a set such that $|A| = n$. Since there exists no onto function $f: A \to \mathcal{P}(A)$, it follows from Theorem 5.7 that if $\mathcal{P}(A)$ is finite then $|\mathcal{P}(A)| > n$. The goal of our next theorem is to count the elements of $\mathcal{P}$(A).

The next set of exercises will help us to do this. We recall that the letter $\theta$ is called "theta," and that $\psi$ is called "psi."

**Exercises (5.9)** Let $A = \{7, 11, 13\}$. Let $\mathcal{G} = \{G \in \mathcal{P}(A) \mid 7 \notin G\}$ and let $\mathcal{H} = \{H \in \mathcal{P}(A) \mid 7 \in H\}$. Let $\theta: \mathcal{G} \to \mathcal{H}$ be defined by $\theta(G) = G \cup \{7\}$.

**1.** List the elements of $\mathcal{G}$.

**2.** List the elements of $\mathcal{H}$.

**3.** Let $\mathcal{S} = \{\varnothing, \{11, 13\}\}$. Find $\theta(\mathcal{S})$.

**4.** Let $\mathcal{T} = \{\{7, 11\}, \{7\}\}$. Find $\theta^{-1}(\mathcal{T})$.

5. Theorem. Let $A$, $B$ be sets such that $A \subseteq B$. For each $x \in B$, $(A - \{x\}) \cup \{x\} = A$ if and only if $x \in A$.

6. Theorem. Let $A$, $B$ be sets such that $A \subseteq B$. For each $x \in B$, $(A \cup \{x\}) - \{x\} = A$ if and only if $x \notin A$.

7. Theorem. Let $A$ be a nonempty set, and let $x \in A$. Let $\mathscr{G} = \{G \in \mathscr{P}(A) \mid x \notin G\}$. Let $\mathscr{H} = \{H \in \mathscr{P}(A) \mid x \in H\}$. Let $\theta: \mathscr{G} \to \mathscr{H}$ be defined by $\theta(G) = G \cup \{x\}$. Let $\psi: \mathscr{H} \to \mathscr{G}$ be defined by $\psi(H) = H - \{x\}$. Then $\psi$ is the inverse of $\theta$, and both $\theta$ and $\psi$ are bijections.

**Remark.** Our next theorem is proved by induction on $\omega_0$, rather than induction on $\mathbb{N}$. In Exercise 5.4.6, the reader proved that for each $m \in \mathbb{Z}$, the principle of mathematical induction applies to the set $\{x \in \mathbb{Z} \mid x \geq m\}$. Since $\omega_0 = \{x \in \mathbb{Z} \mid x \geq 0\}$, we can use induction to prove theorems about $\omega_0$.

**Theorem (5.11)** For each $k \in \omega_0$, for each set $A$, if $|A| = k$ then $|\mathscr{P}(A)| = 2^k$.

**Proof.** The proof is by induction on $k$.

Let $k = 0$. Let $A$ be a set such that $|A| = 0$. Then $A = \varnothing$. Since $A = \varnothing$, $A$ has only one subset (namely, $A$ itself). Thus $\mathscr{P}(A)$ has exactly one element. Since $|\mathscr{P}(A)| = 1$ and $2^0 = 1$, $|\mathscr{P}(A)| = 2^0$.

Let $k \in \omega_0$, and suppose that for each set $A$, if $|A| = k$ then $|\mathscr{P}(A)| = 2^k$.

Let $B$ be a set such that $|B| = k + 1$. We will show that $|\mathscr{P}(B)| = 2^{k+1}$. Let $x \in B$. Let $\mathscr{G} = \{G \in \mathscr{P}(B) \mid x \notin G\}$. Then $\mathscr{G} = \mathscr{P}(B - \{x\})$. Since $|B - \{x\}| = k$, by the induction hypothesis $|\mathscr{P}(B - \{x\})| = 2^k$. Since $\mathscr{G} = \mathscr{P}(B - \{x\})$, $|\mathscr{G}| = 2^k$.

Let $\mathscr{H} = \{H \in \mathscr{P}(B) \mid x \in B\}$. Since, by Exercise 7, there exists a bijection $\theta: \mathscr{G} \to \mathscr{H}$, it follows that $|\mathscr{G}| = |\mathscr{H}|$. Thus $|\mathscr{H}| = 2^k$. Since $\mathscr{P}(B) = \mathscr{G} \cup \mathscr{H}$ and $\mathscr{G} \cap \mathscr{H} = \varnothing$, $|\mathscr{P}(B)| = |\mathscr{G}| + |\mathscr{H}| = 2^k + 2^k = 2^{k+1}$.

Therefore, by the principle of mathematical induction, for each $k \in \omega_0$ and for each set $A$, if $|A| = k$ then $|\mathscr{P}(A)| = 2^k$. Q.E.D.

**Remark.** Induction is not only a method of proof, but also a way of defining functions, sequences, and other mathematical objects. Let $A$ be a set. To define a function $f: \mathbb{N} \to A$ inductively, first, for some $k \in \mathbb{N}$, we define $f(m)$ for all $m \in \mathbb{N}_k$. Then, for each $n \geq k$, we define $f(n+1)$ in terms of $f(m)$, for $m \in \mathbb{N}_n$. Inductive definition is also known as *recursive* definition, and functions defined inductively are also called *recursive* functions.

**Example (5.4)** The *Fibonacci numbers* $\{F_n\}_{n=1}^{\infty}$ are defined as follows. Set $F_1 = F_2 = 1$, and for all $n \in \mathbb{N}$, $F_{n+2} = F_n + F_{n+1}$.

**Remark.** Let $f: \mathbb{N} \to \mathbb{N}$ be defined by $f(n) = F_n$. Then $f$ is a recursive function.

**Exercises (5.10)**

1. Find the first twenty Fibonacci numbers.

2. Theorem. Let $T \subseteq \mathbb{N}$. Suppose there exists $k \in \mathbb{N}$ such that $\mathbb{N}_k \subseteq T$ and for all $m \geq k$, if $\mathbb{N}_m \subseteq T$ then $\mathbb{N}_{m+1} \subseteq T$. Then $T = \mathbb{N}$.

   The theorem of Exercise 2 states a form of the principle of induction where the base step consists of proving that each element of $\mathbb{N}_k$ is in $T$, for some $k \in \mathbb{N}$. To prove the theorem of Exercise 3, we use the theorem of Exercise 2 with $k = 2$. Thus, we use induction with two base steps, the cases $n = 1$ and $n = 2$. Proving the theorem of Exercise 4 by induction requires 4 base steps.

3. Theorem. For each $n \in \mathbb{N}$, let $F_n$ be the $n$th Fibonacci number. Then for each $n \in \mathbb{N}$, $F_n$ is even if and only if $3|n$.

4. Theorem. For each $n \in \mathbb{N}$, let $F_n$ be the $n$th Fibonacci number. Then for each $n \in \mathbb{N}$, $3|F_n$ if and only if $4|n$.

5. Theorem. For each $n \in \mathbb{N}$, let $F_n$ be the $n$th Fibonacci number. Then for each $n \in \mathbb{N}$, $(F_n F_{n+2}) - F_{n+1}^2 = (-1)^{n+1}$.

6. Theorem. For each $n \in \mathbb{N}$, let $F_n$ be the $n$th Fibonacci number. Then for each $n \in \mathbb{N}$, $F_n F_{n+3} - F_{n+1} F_{n+2} = (-1)^{n+1}$.

7. Theorem. For each $n \in \mathbb{N}$, let $F_n$ be the $n$th Fibonacci number. Then for each $n$, $k \in \mathbb{N}$, $F_{n+k+1} = F_{k+1} F_{n+1} + F_k F_n$. [Hint: Use induction on $k$.]

8. Theorem. Let $A$, $B$ be sets and let $f: A \to B$. If $f$ is one-to-one, then $|f(A)| = |A|$.

9. Theorem. Let $A$ be a set, let $n \in \mathbb{N}$, and let $f: \mathbb{N} \to A$. If $f|\mathbb{N}_n$ is one-to-one and $f(n+1) \notin f(\mathbb{N}_n)$, then $f|\mathbb{N}_{n+1}$ is one-to-one.

10. Theorem. Let $A$ be a set and let $f: \mathbb{N} \to A$ be a function. If for each $n \in \mathbb{N}$, $f|\mathbb{N}_n$ is one-to-one, then $f$ is one-to-one.

11. Theorem. Let $A$, $B$ be sets and let $f: A \to B$ be one-to-one. For all $C \subseteq A$, if $C \neq A$ then $f(C) \neq f(A)$.

12. Theorem. Let $A$ and $B$ be sets such that $A \subseteq B$, $B$ is infinite, and $A$ is finite. Then $B - A$ is infinite.

13. Theorem. Let $A$, $B$ be sets, and let $g: B \to A$ be one-to-one. Let $\tilde{g}: B \to \text{Im } g$ be defined by $\tilde{g}(x) = g(x)$. Since $\tilde{g}: B \to \text{Im } g$ is a bijection, there exists an inverse bijection $\tilde{g}^{-1}: \text{Im } g \to B$. Let $h: B \to B$ be one-to-one but not onto. Let $f: A \to A$ be defined by

$$f(x) = \begin{cases} (g \circ h \circ \tilde{g}^{-1})(x), \text{ if } x \in \text{Im } g; \\ x \text{ otherwise.} \end{cases}$$

Then $f$ is one-to-one but not onto. [Hints: Notice that $f(\text{Im } g) \subseteq \text{Im } g$ and $f(A - \text{Im } g) \subseteq A - \text{Im } g$. It may be helpful to draw a diagram with arrows representing the various functions.]

Our next theorem concerns infinite sets.

**Theorem (5.12)** Let $A$ be an infinite set. Then there exists a one-to-one function $f: \mathbb{N} \to A$.

**Proof.** Let $A$ be an infinite set. Since $A$ is infinite and $\varnothing$ is finite, $A \neq \varnothing$. Since $A \neq \varnothing$, there exists $x \in A$. Let $f(1) = x$.

Let $n \in \mathbb{N}$ and suppose that for each $k \in \mathbb{N}_n$, $f(k)$ has already been defined, and that $f|\mathbb{N}_n$ is one-to-one. We define $f(n+1)$ as follows. Since $f|\mathbb{N}_n$ is one-to-one, $f|\mathbb{N}_n$ is a bijection from $\mathbb{N}_n$ to $f(\mathbb{N}_n)$. Since such a bijection exists, $f(\mathbb{N}_n)$ is finite. Since $f(\mathbb{N}_n)$ is finite and $A$ is infinite, $f(\mathbb{N}_n) \neq A$. Since $f(\mathbb{N}_n) \neq A$ and $f(\mathbb{N}_n) \subseteq A$, $A \not\subseteq f(\mathbb{N}_n)$. Therefore, there exists $y \in A - f(\mathbb{N}_n)$. Let $f(n+1) = y$.

Since $f|\mathbb{N}_n$ is one-to-one and $f(n+1) \notin f(\mathbb{N}_n)$, by Exercise 9 $f|\mathbb{N}_{n+1}$ is one-to-one.

Thus we have recursively defined a function $f: \mathbb{N} \to A$ such that for each $n \in \mathbb{N}$, $f|\mathbb{N}_n$ is one-to-one

Hence, by Exercise 10, $f$ is one-to-one.

Therefore, for each set $A$, if $A$ is infinite then there exists a one-to-one function $f: \mathbb{N} \to A$. Q.E.D.

**Theorem (5.13)** Let $A$ be a set. If $A$ is infinite then there exists $f: A \to A$ such that $f$ is one-to-one but not onto.

**Proof.** Let $A$ be an infinite set. Since $A$ is infinite, there exists a one-to-one function $g: \mathbb{N} \to A$. Since $g$ is one-to-one, $\tilde{g}: \mathbb{N} \to \operatorname{Im} g$ is a bijection. Therefore, there exists an inverse function $\tilde{g}^{-1}: \operatorname{Im} g \to \mathbb{N}$. Since $\tilde{g}^{-1}$ is invertible, $\tilde{g}^{-1}$ is a bijection. Let $h: \mathbb{N} \to \mathbb{N}$ be defined by $h(x) = 2x$. Then $h$ is one-to-one but not onto. Let $f: A \to A$ be defined by

$$f(x) = \begin{cases} (g \circ h \circ \tilde{g}^{-1})(x), \text{ if } x \in \ Im\ g \\ x \text{ otherwise.} \end{cases}$$

By Exercise 13, $f$ is one-to-one but not onto.

Therefore, for each set $A$, if $A$ is infinite then there exists $f: A \to A$ such that $f$ is one-to-one but not onto. Q.E.D.

Exercises (5.11)

1. Theorem. For each set $A$, $A$ is finite if and only if for each $f: A \to A$, $f$ is one-to-one if and only if $f$ is onto.

2. Theorem. Let $A$ and $B$ be sets and let $f: A \to B$. If $A$ is finite, then $f(A)$ is finite.

3. Theorem. Let $A$ and $B$ be sets and let $f: A \to B$. If $A$ is finite and $f$ is not one-to-one, then $|f(A)| < |A|$.

4. Theorem. For each set $A$, for each $C \subseteq A$, if $A$ is finite then $C$ is finite.

   Corollary. For each set $A$, for each $C \subseteq A$, if C is infinite then $A$ is infinite.

5. False Proposition. For all sets $A$, $B$, if $B \subseteq A$ and B is infinite, then $A - B$ is finite.

6. Theorem. For all sets $A$, $B$, for each $f: A \to B$, if $f$ is onto and $B$ is infinite, then $A$ is infinite.

7. Theorem. For all sets $A$, $B$, for each $f: A \to B$, if $f$ is one-to-one and $A$ is infinite, then $B$ is infinite.

The following exercises will help us prove the next major theorem.

8. Theorem. Let $X$ be an infinite subset of $\mathbb{N}$. Define $f: \mathbb{N} \to X$ as follows. Let $f(1)$ be the least element of $X$. Let $n \in \mathbb{N}$, and assume that for each $m \in \mathbb{N}_n$, $f(m)$ has already been defined. Since $X$ is infinite and $f(\mathbb{N}_n)$ is finite, $X - f(\mathbb{N}_n) \neq \varnothing$. Hence, by the well-ordering property of the natural numbers, $X - f(\mathbb{N}_n)$ has a least element. Let $f(n+1)$ be the least element of $X - f(\mathbb{N}_n)$.

   For all $a$, $b \in \mathbb{N}$, if $a < b$ then $f(a) \neq f(b)$.

9. Theorem. Let $X$ be an infinite subset of $\mathbb{N}$. Define $f: \mathbb{N} \to X$ as follows. Let $f(1)$ be the least element of $X$. Let $n \in \mathbb{N}$, and assume that for each $m \in \mathbb{N}_n$, $f(m)$ has already been defined. Let $f(n+1)$ be the least element of $X - f(\mathbb{N}_n)$.

   Then for each $k \in \mathbb{N}$, if $\mathbb{N}_k \cap X \neq \varnothing$ then there exists $r \in \mathbb{N}$ such that $f(\mathbb{N}_r) = \mathbb{N}_k \cap X$. [Hint: Use induction on $k$.]

**Theorem (5.14)** Let $X$ be an infinite subset of $\mathbb{N}$. Then $|\mathbb{N}| = |X|$.

**Proof.** We define a function $f: \mathbb{N} \to X$ recursively as follows. Let $f(1)$ be the least element of $X$. Let $n \in \mathbb{N}$, and suppose that for each $m \in \mathbb{N}_n$, $f(m)$ has already been defined. Since $f(\mathbb{N}_n)$ is finite and $X$ is infinite, $X - f(\mathbb{N}_n) \neq \varnothing$. Hence, by the well-ordering property of the natural numbers, $X - f(\mathbb{N}_n)$ has a least element. Let $f(n+1)$ be the least element of $X - f(\mathbb{N}_n)$.

We claim that $f: \mathbb{N} \to X$ is a bijection.

First we will show that $f$ is one-to-one. Let $a$, $b \in \mathbb{N}$ such that $f(a) = f(b)$. By Exercise 8, $b \leq a$. But also, since $f(b) = f(a)$, it follows from Exercise 8 that $a \leq b$. Since $b \leq a$ and $a \leq b$, $a = b$.

Thus for all $a, b \in \mathbb{N}$, if $f(a) = f(b)$ then $a = b$. That is, $f$ is one-to-one.

It remains to show that $f$ is onto. Let $x \in X$. Either $x$ is the least element of $X$, or not.

Case 1. Suppose that $x$ is the least element of $X$. Then $f(1) = x$. Hence $x \in \operatorname{Im} f$.

Case 2. Suppose that $x$ is not the least element of $X$. Then $x \neq 1$. Since $x \in \mathbb{N}$ and $x \neq 1$, $x - 1 \in \mathbb{N}$. Since $x$ is not the least element of $X$, $X \cap \mathbb{N}_{x-1} \neq \varnothing$. By Exercise 9, since $X \cap \mathbb{N}_{x-1} \neq \varnothing$, there exists $r \in \mathbb{N}$ such that $f(\mathbb{N}_r) = X \cap \mathbb{N}_{x-1}$. Since $x$ is the least element of $X - (X \cap \mathbb{N}_{x-1})$, it follows that $x$ is the least element of $X - f(\mathbb{N}_r)$. Therefore, by definition of $f$, $f(r+1) = x$. Hence $x \in \operatorname{Im} f$.

Thus, for all $x \in X$, $x \in \text{Im } f$. Hence for each $x \in X$ there exists $k \in \mathbb{N}$ such that $f(k) = x$. That is, $f$ is onto.

Since $f$ is both one-to-one and onto, $f$ is a bijection. Since there exists a bijection $f: \mathbb{N} \to X$, it follows that $|\mathbb{N}| = |X|$.

Therefore, for each infinite subset $X$ of $\mathbb{N}$, $|\mathbb{N}| = |X|$. Q.E.D.

**Definition.** Let $A$ be a set. Then $A$ is *countably infinite* if $|\mathbb{N}| = |A|$. If $A$ is finite or $A$ is countably infinite, then $A$ is *countable*. Otherwise, $A$ is *uncountable*.

**Exercises (5.12)**

1. Theorem. The set $\mathbb{Z}$ of integers is countably infinite.

2. Theorem. Let $A$ be a set. If there exist functions $f: A \to \mathbb{N}$ and $g: A \to \mathbb{N}$ such that $f$ is one-to-one and $g$ is onto, then $|A| = |\mathbb{N}|$.

3. Theorem. Let $f: \mathbb{N} \times \mathbb{N} \to \mathbb{N}$ be defined: for each $m, n \in \mathbb{N}$, $f((m, n)) = 2^m 3^n$. Then $f$ is one-to-one.

4. Theorem. The set $\mathbb{N} \times \mathbb{N}$ is countably infinite. [Hint: Exercise 3 and Theorem 5.14 give a simple proof of this theorem.]

   In Exercise 4, we proved that there exists a bijection $f: \mathbb{N} \times \mathbb{N} \to \mathbb{N}$, but we did not construct $f$ explicitly. In Exercises 5 – 10, we will actually construct a bijection $f$ from $\mathbb{N} \times \mathbb{N}$ to $\mathbb{N}$.

5. Theorem. Let $A$ be a set, let $\mathscr{S} = \{X_k\}_{k=1}^{\infty}$ be a partition of $A$, and let $f: A \to B$. If, for each $k \in \mathbb{N}$, $f|X_k$ is one-to-one and if $\{f(X_k)\}_{k=1}^{\infty}$ is a partition of $B$, then $f$ is a bijection.

6. Theorem. For each $k \in \mathbb{N}$, let $X_k \subseteq \mathbb{N} \times \mathbb{N}$ be the set $(\{k\} \times \mathbb{N}_k) \cup (\mathbb{N}_k \times \{k\})$. Let $\mathscr{S} = \{X_k\}_{k=1}^{\infty}$. Then $\mathscr{S}$ is a partition of $\mathbb{N} \times \mathbb{N}$.

7. Theorem. For each $k \in \mathbb{N}$, let $X_k \subseteq \mathbb{N} \times \mathbb{N}$ be the set $(\{k\} \times \mathbb{N}_k) \cup (\mathbb{N}_k \times \{k\})$. Let $f: X_k \to \mathbb{N}$ be defined by

$$f(m, n) = \begin{cases} m + (n-1)^2, \text{ if } n = k; \\ n + m + (m-1)^2 \text{ otherwise.} \end{cases}$$

Then $f$ is one-to-one, and $\text{Im } f = \{x \in \mathbb{N} \mid (k-1)^2 < x \leq k^2\}$.

8. Theorem. For each $k \in \mathbb{N}$, let $T_k \subseteq \mathbb{N}$ be the set $\{x \in \mathbb{N} \mid (k-1)^2 < x \leq k^2\}$. Let $\mathscr{T} = \{T_k\}_{k=1}^{\infty}$. Then $\mathscr{T}$ is a partition of $\mathbb{N}$.

9. Theorem. Let $f: \mathbb{N} \times \mathbb{N} \to \mathbb{N}$ be defined by

$$f((m, n)) = \begin{cases} m + (n-1)^2, \text{ if } m \leq n; \\ n + m + (m-1)^2 \text{ otherwise.} \end{cases}$$

Then $f$ is a bijection.

10. List the elements of $\mathbb{N} \times \mathbb{N}$ such that $m, n \leq 7$ in a rectangular array, as shown below.

$$\begin{array}{ccccccc} (1,1) & (1,2) & (1,3) & (1,4) & (1,5) & (1,6) & (1,7) \\ (2,1) & (2,2) & (2,3) & (2,4) & (2,5) & (2,6) & (2,7) \\ (3,1) & (3,2) & (3,3) & (3,4) & (3,5) & (3,6) & (3,7) \\ (4,1) & (4,2) & (4,3) & (4,4) & (4,5) & (4,6) & (4,7) \\ (5,1) & (5,2) & (5,3) & (5,4) & (5,5) & (5,6) & (5,7) \\ (6,1) & (6,2) & (6,3) & (6,4) & (6,5) & (6,6) & (6,7) \\ (7,1) & (7,2) & (7,3) & (7,4) & (7,5) & (7,6) & (7,7) \end{array}$$

Let $f: \mathbb{N} \times \mathbb{N} \to \mathbb{N}$ be defined by

$$f(m, n) = \begin{cases} m + (n-1)^2, \text{ if } m \leq n; \\ m + n + (m-1)^2, \text{ otherwise.} \end{cases}$$

For each $(m, n) \in \mathbb{N}_7 \times \mathbb{N}_7$, write the number $f((m, n))$ to the left of $(m, n)$ in your array. This will give us a picture of the function $f$.

11. Theorem. Let $A \subseteq \mathbb{Q}$ be the set $\{x \in \mathbb{Q} \mid x > 0\}$. Then $|\mathbb{Q}| = |A|$.

12. Theorem. The set $\mathbb{Q}$ is countably infinite.

**Remark.** It may seem as if, given any infinite set $A$, we can prove that $A$ is countably infinite. This is definitely not true. Recall that we have proved that for any set $A$ and for any function $f: A \to \mathcal{P}(A)$, the function $f$ fails to be onto. Since there is no surjection from a set to its own power set, there is no bijection from a set to its own power set. Therefore, for each set $A$, $|A| \neq |\mathcal{P}(A)|$.

Consider the set $\mathcal{P}(\mathbb{N})$, the power set of the natural numbers. Since $\mathcal{P}(\mathbb{N})$ has infinitely many elements and $|\mathcal{P}(\mathbb{N})| \neq |\mathbb{N}|$, it follows that $\mathcal{P}(\mathbb{N})$ is uncountable.

We have proved that the sets $\mathbb{Z}$, $\mathbb{Q}$, and $\mathbb{N} \times \mathbb{N}$ are countable. What about the set $\mathbb{R}$ of real numbers?

**Real numbers.** The full glory of the set $\mathbb{R}$ of real numbers is the subject of mathematics courses in real analysis, and lies beyond the scope of this book. We do not assume knowledge of analysis or calculus on the part of the reader. Still, we must say *something* about the real numbers, if we wish to find out whether or not the set $\mathbb{R}$ is countable.

**Notation.** Let $a, b \in \mathbb{R}$ such that $a < b$. The open interval $(a, b)$ is the set $\{x \in \mathbb{R} \mid a < x < b\}$.

**Remark.** Of course, the open interval $(a, b) \subseteq \mathbb{R}$ is not the same as the ordered pair $(a, b) \in \mathbb{R}^2$. It may seem confusing to use the notation $(a, b)$ to mean two different things, but in practice it is no worse than having two different meanings for the word (or string of letters) "rose" in English. In each particular context, only one meaning makes sense.

Let us consider the open interval $(0, 1)$ in $\mathbb{R}$.

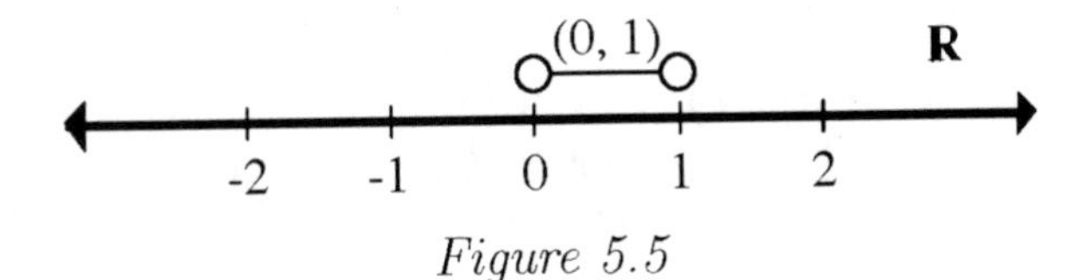


*Figure 5.5*

The open interval (0, 1) contains all the numbers that are larger than 0 and smaller than 1. Some of the these numbers are rational (for example, $\frac{1}{2}, \frac{3}{7}, \frac{11}{17}, \frac{298}{303}$) and some are irrational. An irrational number is a real number which cannot be expressed as a fraction with integer numerator and integer denominator. For example, $\frac{1}{\sqrt{2}}$ is irrational. (The symbol $\frac{1}{\sqrt{2}}$ may look like a fraction, but its denominator is not an integer.) Here is a simple proof that this number is irrational.

**Theorem (5.15)** The number $\frac{1}{\sqrt{2}}$ is irrational.

**Proof.** By way of contradiction, suppose that $\frac{1}{\sqrt{2}} \in \mathbb{Q}$. Then there exist $p \in \mathbb{Z}$ and $q \in \mathbb{N}$ such that $\frac{1}{\sqrt{2}} = \frac{p}{q}$ and $p$ and $q$ have no common factors greater than 1. Since $\frac{1}{\sqrt{2}} = \frac{p}{q}$, it follows that $q = p\sqrt{2}$ and thus that $q^2 = 2p^2$. Since $2 \mid q^2$, it follows that $2 \mid q$. Since $2 \mid q$, we know that $4 \mid q^2$. Since $4 \mid 2p^2$, $2 \mid p^2$. Since $2 \mid p^2$,the number $p$ must be even. So $p$ and $q$ are both divisible by 2. But by hypothesis $p$ and $q$ have no common factors greater than 1. $\rightarrow\leftarrow$

Our hypothesis has led to a contradiction and is therefore false. Therefore, the number $\frac{1}{\sqrt{2}}$ is irrational. Q.E.D.

The foregoing theorem implies that some of the numbers in (0, 1) are irrational.

How can we describe the elements of the open interval (0, 1), given that we do not know much about them? What we do know is that each number in (0, 1) has a decimal expansion. That is, for each $x \in (0, 1)$, the number $x$ can be written (in principle) as a zero followed by a decimal point followed by an infinite string of digits. (In practice, nobody can write down an infinite string of digits.)

**Notation.** Let $x \in (0, 1)$. Then for each $k \in \mathbb{N}$, there exists $a_k \in \omega_0$ such that $a_k \leq 9$ and the number $x$ can be written as a zero followed by a decimal point followed by the digits $a_1 a_2 a_3 a_4 \ldots$ Then we write

$x = \sum_{k=1}^{\infty} 10^{-k}a_k$. The sum $\sum_{k=1}^{\infty} 10^{-k}a_k$ is called a *decimal expansion* of the number $x$.

**Remark.** Readers familiar with calculus or analysis are aware that the symbol $\sum_{k=1}^{\infty} 10^{-k}a_k$ denotes $\lim_{n\to\infty} \sum_{k=1}^{n} 10^{-k}a_k$. Since $\lim_{n\to\infty} \sum_{k=1}^{n} 10^{-k}a_k$ is the number whose decimal expansion has $a_k$ as its $k$th digit after the decimal point, the notation we are using is correct. But we will not discuss limits in the main text of this book. We say a few words about limits in Afterword B.

**Notation.** The number $0.777777\ldots$ is often written $0.\overline{7}$. Similarly the number $0.2937373737\ldots$ may be written $0.29\overline{37}$. The bar over a string of digits means that the same string is repeated over and over without end.

**Examples (5.5)**

1. Let $x = \frac{1}{7}$. Then $x = 0.\overline{142857}$. Thus $x = (0.1 \times 1) + (0.01 \times 4) + (0.001 \times 2) + \ldots$ That is, $x = (10^{-1} \times 1) + (10^{-2} \times 4) + (10^{-3} \times 2) + \ldots$ Thus $x = \sum_{k=1}^{\infty} 10^{-k}a_k$, where $a_1 = 1$, $a_2 = 4$, $a_3 = 2$, $a_4 = 8$, $a_5 = 5$, $a_6 = 7$, and for all $k \geq 7$, $a_k = a_{k-6}$. The decimal expansion of $\frac{1}{7}$ repeats itself after every 6 digits.

2. Let $x = \frac{1}{\sqrt{2}}$. Then $x = 0.70710678\ldots$ Thus $x = \sum_{k=1}^{\infty} 10^{-k}a_k$, where $a_1 = 7$, $a_2 = 0$, $a_3 = 7\ldots$ Since $\frac{1}{\sqrt{2}}$ is irrational, its decimal expansion does not repeat itself.

3. Let $x = 9\sum_{k=1}^{\infty} 10^{-k}$. Then $x = 0.\overline{9}$. We will show that $x = 1$. Since $x = 0.\overline{9}$, $10x = 9.\overline{9}$. Thus $10x - x = 9.\overline{9} - \overline{9} = 9$. Since $9x = 9$, $x = 1$.

4. Let $x = \frac{1}{2}$. Then $x = 0.5$. That is, $x = 0.5\overline{0}$. Thus $x = \sum_{k=1}^{\infty} 10^{-k}a_k$, where $a_1 = 5$ and for all $n \geq 2$, $a_k = 0$. But also $x = 0.4\overline{9}$. This

means that $x = \sum_{k=1}^{\infty} 10^{-k} b_k$, where $b_1 = 4$ and for all $k \geq 2$, $b_k = 9$. Thus the number $\frac{1}{2}$ has two different decimal expansions.

**Theorem (5.16)** Let $x \in \mathbb{R}$, and let $\sum_{k=1}^{\infty} 10^{-k} a_k$ be a decimal expansion of $x$. If there exists $m \in \mathbb{N}$ such that $a_m \neq 0$, then $x > 0$. If there exists $n \in \mathbb{N}$ such that $a_n \neq 9$, then $x < 1$.

**Proof.** Let $x \in \mathbb{R}$, and let $\sum_{k=1}^{\infty} 10^{-k} a_k$ be a decimal expansion of $x$. Suppose that there exists $m \in \mathbb{N}$ such that $a_m \neq 0$. Then $x \geq 10^{-m} a_m$. Hence $x > 0$. Thus, if there exists $m \in \mathbb{N}$ such that $a_m \neq 0$, then $x > 0$.

Suppose that there exists $n \in \mathbb{N}$ such that $a_n \neq 9$. Since $1 = 9 \sum_{k=1}^{\infty} 10^{-k}$ and $a_n \neq 9$, $1 - x \geq 10^{-n}(9 - a_n)$. Since $a_n \leq 9$ and $a_n \neq 9$, $9 - a_n > 0$. Thus $10^{-n}(9 - a_n) > 0$. Since $1 - x > 0$, $x \neq 1$. Thus, if there exists $n \in \mathbb{N}$ such that $a_n \neq 9$, then $x < 1$.

Therefore, for each $x \in \mathbb{R}$, if $x$ has the decimal expansion $\sum_{k=1}^{\infty} 10^{-k} a_k$ and there exists $m \in \mathbb{N}$ such that $a_m \neq 0$, then $x > 0$. If there exists $n \in \mathbb{N}$ such that $a_n \neq 9$, then $x < 1$. Q.E.D.

**Remark.** In order to prove that the open interval $(0, 1) \in \mathbb{R}$ is uncountable, we need to prove a preliminary "technical lemma." A technical lemma is an easy but awkward theorem that is used to prove another theorem. In this case the technical lemma states that for each $x \in (0, 1)$, if $x$ has two different decimal expansions, then one expansion ends all in 0's and the other all in 9's. Since each real number has at least one decimal expansion, this technical lemma implies that for each $x \in (0, 1)$, $x$ has a unique decimal expansion that does not end all in 9's.

This lemma is often proved by handwaving. We invite the reader to prove it in the next set of exercises. The exercises are quite simple, once you get used to the notation of decimal expansions.

Exercises (5.13)

1. Theorem. Let $x \in (0, 1)$, and let $\sum_{k=1}^{\infty} 10^{-k} a_k$ be a decimal expansion of $x$. For each $p \in \mathbb{N}$, $x \geq \sum_{k=1}^{p} 10^{-k} a_k$.

2. Theorem. Let $x \in (0, 1)$, and let $\sum_{k=1}^{\infty} 10^{-k} b_k$ be a decimal expansion of $x$. For each $p \in \mathbb{N}$, if $b_p < 9$ then $x \leq \sum_{k=1}^{p-1} 10^{-k} b_k + 10^{-p}(b_p + 1)$.

3. Theorem. For each $k \in \mathbb{N}$, let $b_k \in \omega_0$ such that $b_k \leq 9$. Then for each $p \in \mathbb{N}$, if $b_p < 9$ then $\sum_{k=1}^{p} 10^{-k} b_k + \sum_{k=p+1}^{\infty} 10^{-k}(9) = \sum_{k=1}^{p-1} 10^{-k} b_k + 10^{-p}(b_p + 1)$.

4. Theorem. For each $k \in \mathbb{N}$, let $b_k \in \omega_0$ such that $b_k \leq 9$. Then for each $p \in \mathbb{N}$, if $b_p < 9$ and if there exists $n \geq p + 1$ such that $b_n \neq 9$, then $\sum_{k=1}^{\infty} 10^{-k} b_k < \sum_{k=1}^{p-1} 10^{-k} b_k + 10^{-p}(b_p + 1)$.

5. Theorem. For each $k \in \mathbb{N}$, let $a_k \in \omega_0$ such that $a_k \leq 9$. Then for each $p \in \mathbb{N}$, if there exists $m \geq p + 1$ such that $a_m \neq 0$, then $\sum_{k=1}^{\infty} 10^{-k} a_k > \sum_{k=1}^{p} 10^{-k} a_k$.

6. Theorem. Let $x \in (0, 1)$, and let $\sum_{k=1}^{\infty} 10^{-k} a_k$ and $\sum_{k=1}^{\infty} 10^{-k} b_k$ be decimal expansions of $x$. Suppose that there exists $p \in \mathbb{N}$ such that $a_p > b_p$ and for all $k \leq p - 1$, $a_k = b_k$. Then $b_p = a_p - 1$, and for all $k \geq p + 1$, $a_k = 0$ and $b_k = 9$.

7. Theorem. Let $x \in (0, 1)$. Then there exists a unique decimal expansion $\sum_{k=1}^{\infty} 10^{-k} a_k$ of $x$ such that for each $k \in \mathbb{N}$, there exists $n \in \mathbb{N}$ such that $n \geq k$ and $a_n \neq 9$.

**Theorem (5.17)** The open interval (0, 1) is uncountable.

**Proof.** By way of contradiction, suppose that the open interval (0, 1) is countable. Then there exists a bijection $f: \mathbb{N} \to (0, 1)$. Since for each $n \in \mathbb{N}$, $f(n) \in (0, 1)$, there exists a unique decimal expansion $f(n) = \sum_{k=1}^{\infty} 10^{-k} a_{n,k}$ such that for each $k \in \mathbb{N}$, there exists $m \in \mathbb{N}$ such that $m \geq k$ and $a_{n,m} \neq 9$.

For each $k \in \mathbb{N}$, let $b_k = \begin{cases} 7, \text{ if } a_{k,k} = 3; \\ 3 \text{ otherwise.} \end{cases}$

Since for each $k \in \mathbb{N}$, $b_k \in \omega_0$ and $b_k \leq 9$, $\sum_{k=1}^{\infty} 10^{-k} b_k$ is a decimal expansion for a real number. Let $y \in \mathbb{R}$ be the number such that $y = \sum_{k=1}^{\infty} 10^{-k} b_k$. Since $b_1 \neq 0$ and $b_1 \neq 9$, $y > 0$ and $y < 1$. Thus $y \in (0, 1)$.

Since $f$ is a bijection, $f$ is onto. Since $f$ is onto and $y \in (0, 1)$, there exists $p \in \mathbb{N}$ such that $f(p) = y$. Since for each $k \in \mathbb{N}$, $b_k \in \{3, 7\}$, it follows that for each $k \in \mathbb{N}$, $b_k \neq 9$. Thus $y = \sum_{k=1}^{\infty} 10^{-k} a_{p,k}$ and $y = \sum_{k=1}^{\infty} 10^{-k} b_k$ and for each $k \in \mathbb{N}$, there exist $m, r \in \mathbb{N}$ such that $a_{p,m} \neq 9$ and $b_r \neq 9$. Therefore, by Exercise 7, for each $k \in \mathbb{N}$, $a_{p,k} = b_k$. Hence $a_{p,p} = b_p$. Either $a_{p,p} = 3$ or $a_{p,p} \neq 3$.

Case 1. Suppose $a_{p,p} = 3$. Then $b_p = 7$. Since $3 \neq 7$, $a_{p,p} \neq b_p$. Thus $a_{p,p} = b_p$ and $a_{p,p} \neq b_p$. $\rightarrow\leftarrow$

Case 2. Suppose $a_{p,p} \neq 3$. Then $b_p = 3$. Since $a_{p,p} \neq 3$, and $b_p = 3$, $a_{p,p} \neq b_p$. Thus $a_{p,p} = b_p$ and $a_{p,p} \neq b_p$. $\rightarrow\leftarrow$

Our hypothesis has led to a contradiction and is therefore false. Therefore, the open interval (0, 1) is uncountable. Q.E.D.

**Remarks.** The foregoing proof is a version of Cantor's famous "diagonal argument." By way of contradiction, we suppose that there exists a bijection $f: \mathbb{N} \to (0, 1)$. Thus we can index the set (0, 1) by the elements of $\mathbb{N}$. We can make an infinite list, writing each number in (0, 1) as a zero followed by a decimal point, followed by an infinite string of digits.

$$\begin{array}{lllll}
x_1 &= 0.a_{1,1}a_{1,2}a_{1,3}a_{1,4}a_{1,5} \ldots a_{1,n} \ldots \\
x_2 &= 0.a_{2,1}a_{2,2}a_{2,3}a_{2,4}a_{2,5} \ldots a_{2,n} \ldots \\
\vdots & \quad \vdots \quad\quad \vdots \quad\quad \vdots \\
x_k &= 0.a_{k,1}a_{k,2}a_{k,3}a_{k,4}a_{k,5} \ldots a_{k,n} \ldots \\
\vdots & \quad \vdots \quad\quad \vdots \quad\quad \vdots
\end{array}$$

The "diagonal" digits in this array are $a_{1,1}$, $a_{2,2}$, and so on. Then we construct a sequence of digits $\{b_k\}_{k=1}^{\infty}$ such that for each $k \in \mathbb{N}$, $b_k \neq a_{k,k}$ and there exists $n \in \mathbb{N}$ such that $n \geq k$ and $b_k \neq 9$.

The number $x = 0.b_1b_2b_3 \ldots b_n \ldots$ should appear on the list. So there exists $p \in \mathbb{N}$ such that $f(p) = x$. Hence $a_{p,p} = b_p$. But $b_p$ was chosen purposely so that $a_{p,p} \neq b_p$. $\rightarrow\leftarrow$

Thus, it is not possible to use the set $\mathbb{N}$ to index the set (0, 1). That is, (0, 1) is uncountable.

Cantor's diagonal argument is similar in structure to the proof that, for each set $A$, there exists no onto function $f: A \rightarrow \mathscr{P}(A)$. The idea is simple, and the result — that the open interval (0, 1) is uncountable — was astounding when it was first proved. See Afterword A for more on this subject.

Since $\mathbb{R}$ is uncountable, and since $\mathbb{N} \subseteq \mathbb{R}$, it seems that $\mathbb{R}$ is somehow "bigger" than $\mathbb{N}$, in the sense that $\mathbb{R}$ contains more elements than $\mathbb{N}$. Notice that there is no surjection $f: \mathbb{R} \rightarrow \mathscr{P}(\mathbb{R})$. Thus $\mathscr{P}(\mathbb{R})$ is "bigger" than $\mathbb{R}$, and $\mathscr{P}(\mathscr{P}(\mathbb{R}))$ is "bigger" still. The mind boggles, but there is no contradiction here.

When we say that a set $A$ is "bigger" than a set $B$, we are not always discussing cardinality. For example, there is a sense in which the interval $(-1, 2)$ is larger than the interval $(0, 1)$. However, the intervals $(-1, 2)$ and $(0, 1)$ have the same cardinality. They do not have the same *measure*. Mathematicians compare sizes of sets in many different ways.

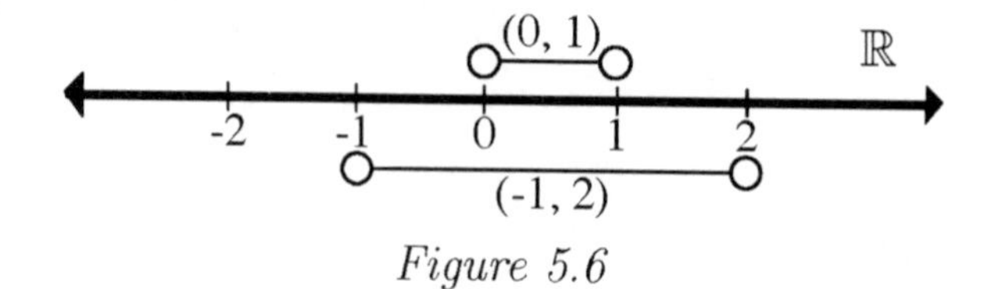

*Figure 5.6*

Exercises (5.14)

1. Theorem. Let $a, b, c, d \in \mathbb{R}$, such that $a < b$ and $c < d$. Let $I$ be the open interval $(a, b)$ and let $J = (c, d)$. Let $f: I \to J$ be defined by $f(x) = \left(\frac{d-c}{b-a}\right) x + \frac{bc-da}{b-a}$. Then $f$ is a bijection.

2. Let $I = (-1, 1)$ and let $J = (2, 3)$. Define $f: I \to J$ as in Exercise 1. Graph the function $f$ on the usual coordinate axes in $\mathbb{R}^2$ as shown below.

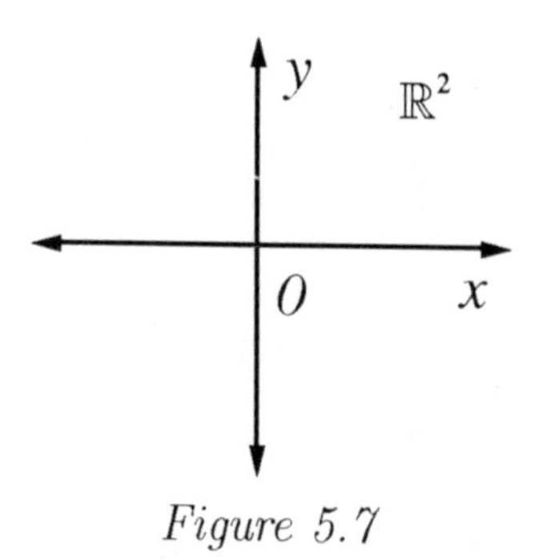

*Figure 5.7*

**3.** Theorem. The set $\mathbb{R}$ and the interval $(-1, 1)$ are equal in cardinality. [Hint: let $f:\mathbb{R}\to(-1, 1)$ be the function defined by

$$f(x) = \begin{cases} \frac{x}{x+1}, & x > 0; \\ \frac{x}{1-x}, & x \leq 0. \end{cases}$$

Show that $f$ is a bijection.]

**4.** Theorem. Let $A$, $B$ be sets. If $A \subseteq B$ and $A$ is uncountable, then $B$ is uncountable.

**5.** Theorem. Let $A$, $B$ be sets. If $A \subseteq B$ and $B$ is uncountable and $A$ is countable, then $B - A$ is uncountable.

**6.** Theorem. The set $\mathbb{R} - \mathbb{Q}$ of irrational numbers is uncountable.

# Chapter 6

# Introduction to Combinatorics

*Murphy receded a little way into the north and prepared to finish his lunch. He took the biscuits carefully out of the packet and laid them face upward on the grass, in order as he felt of edibility. They were the same as always, a Ginger, an Osborne, a Digestive, a Petit Beurre and one anonymous. He always ate the first-named last, because he liked it the best, and the anonymous first, because he thought it very likely the least palatable. The order in which he ate the remaining three was indifferent to him and varied irregularly from day to day. On his knees now before the five it struck him for the first time that these prepossessions reduced to a paltry six the number of ways in which he could make this meal. But this was to violate the very essence of assortment, this was red permanganate on the Rima of variety. Even if he conquered his prejudice against the anonymous, still there would be only twenty-four ways in which the biscuits could be eaten. But were he to take the final step and overcome his infatuation with the ginger, then the assortment would spring to life before him, dancing the radiant measure of its total permutability, edible in a hundred and twenty ways!*

*Overcome by these perspectives Murphy fell forward on his face on the grass, beside those biscuits of which it could be said as truly as of the stars, that one differed from another, but of which he could not partake in their fullness until he had learnt not to prefer any one to any other.*

*Samuel Beckett*, Murphy

In Chapters 6 and 7 we will focus on finite sets. The theory of counting is a fascinating subject that goes by the name of *combinatorics*. In these two chapters we introduce the rudiments of combinatorics.

**Definition.** Let $A$ be a set. Then $A^1 = A$, and for all $n \in \mathbb{N}$, $A^{n+1} = A^n \times A$.

**Pronunciation.** The symbol $A^n$ is pronounced "$A$ $n$," not "$A$ to the $n$th." For example, the symbol $\mathbb{R}^3$ is pronounced "$R$ three."

**Notation.** By definition, $A^2 = A \times A$, the Cartesian product of $A$ with itself. According to the definition of Cartesian product, for each $x \in A^2$, there exist $a_1, a_2 \in A$ such that $x = (a_1, a_2)$. Since $A^3 = A^2 \times A$, for each $x \in A^3$, there exist $a_1, a_2, a_3 \in A$ such that $x = ((a_1, a_2), a_3)$. Since $A^4 = A^3 \times A$, for each $x \in A^4$, there exist $a_1, a_2, a_3, a_4 \in A$ such that $x = (((a_1, a_2), a_3), a_4)$.

We usually write these expressions without inner parentheses. Thus, for all $x \in A^4$ there exist $a_1, a_2, a_3, a_4 \in A$ such that $x = (a_1, a_2, a_3, a_4)$.

**Examples (6.1)**

1. The set $\mathbb{R}^1$ is the number line. The set $\mathbb{R}^2$ is the Cartesian plane. The set $\mathbb{R}^3$ is three dimensional space, or three-space, with three mutually perpendicular axes. The set $\mathbb{R}^4$ is four-dimensional space. And so on.

2. For each $n \in \mathbb{N}$, the set $\mathbb{Z}^n$ is the *integer lattice* in $\mathbb{R}^n$. That is, $\mathbb{Z}^n$ is the set of points in $\mathbb{R}^n$ all of whose coordinates are integers.

3. Let $A = \{0, 1\}$. Then $A^2 = \{(0, 0), (0, 1), (1, 0), (1, 1)\}$, $A^3 = \{(0, 0, 0), (0, 0, 1), (0, 1, 0), (1, 0, 0), (0, 1, 1), (1, 0, 1), (1, 1, 0), (1, 1, 1)\}$, and so on. We can represent the sequence $\{A^n\}_{n \in \mathbb{N}}$ by a tree diagram as follows:

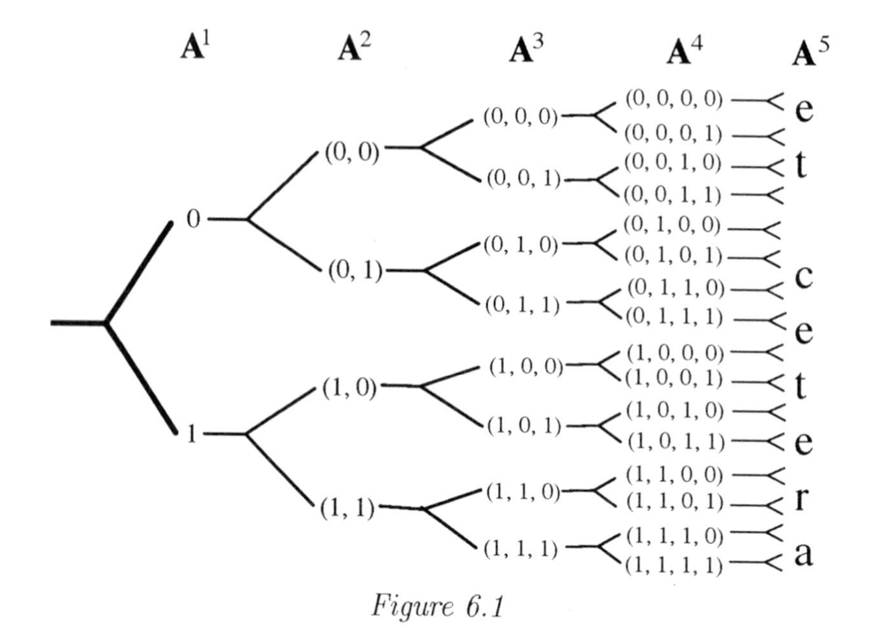


*Figure 6.1*

**Theorem (6.1)** Let $k \in \mathbb{N}$ and let $B$ be a set such that $|B| = k$. Then for all $n \in \mathbb{N}$, for each set $A$, if $|A| = n$ then $|A \times B| = |A||B|$.

**Proof.** The proof is by induction on $n$.

Let $A$ be a set such that $|A| = 1$. Let $a \in A$. Then $A = \{a\}$, and $A \times B = \{(a, b) \mid b \in B\}$. Let $f: A \to B$ be defined by $f(a, b) = b$. Then $f$ is a bijection. (Proof: Exercise.) Thus $|A \times B| = |B| = k$. Therefore, for each set $A$, if $|A| = 1$ then $|A \times B| = |A||B|$.

Let $n \in \mathbb{N}$, and suppose that for each set $A$, if $|A| = n$ then $|A \times B| = |A||B|$. Let $C$ be a set such that $|C| = n + 1$. Let $a \in C$, let $C_1 = \{a\}$ and let $C_2 = C - \{a\}$.

Since $C_1 \cup C_2 = C$, it follows that $C \times B = (C_1 \times B) \cup (C_2 \times B)$. Since $C_1 \cap C_2 = \varnothing$, also $(C_1 \times B) \cap (C_2 \times B) = \varnothing$.

Since $(C_1 \times B) \cap (C_2 \times B) = \varnothing$, it follows that $|(C_1 \times B) \cup (C_2 \times B)| = |C_1 \times B| + |C_2 \times B|$. Since $|C_1| = 1$, $|C_1 \times B| = |C_1||B|$. Since $|C_2| = n$, by the induction hypothesis $|C_2 \times B| = |C_2||B|$. Hence $|C \times B| = |C_1 \times B| + |C_2 \times B| = |C_1||B| + |C_2||B| = (|C_1| + |C_2|)(|B|)$. Since $C_1 \cup C_2 = C$

and $C_1 \cap C_2 = \varnothing$, $|C_1| + |C_2| = |C|$. Hence $(|C_1| + |C_2|)|B| = |C||B|$. Therefore $|C \times B| = |C||B|$. Thus for each set $A$, if $|A| = n + 1$ then $|A \times B| = |A||B|$.

Therefore, by the principle of mathematical induction, for each $n$, $k \in \mathbb{N}$, for all sets $A$, $B$ such that $|A| = n$ and $|B| = k$, $|A \times B| = |A||B|$. Q.E.D.

**Notation.** Let $n \in \mathbb{N}$, and for all $1 \leq i \leq n$, let $a_i \in \mathbb{R}$. The symbol $\prod_{i=1}^{n} a_i$ is shorthand for the product $a_1 a_2 a_3 \ldots a_n$.

**Notation.** Let $n \in \mathbb{N}$, and for each $1 \leq i \leq n$, let $A_i$ be a set. The symbol $\prod_{i=1}^{n} A_i$ represents the Cartesian product $A_1 \times A_2 \times \ldots \times A_n$.

**Remark.** Just as the capital sigma $\Sigma$ stands for "sum," the capital pi $\Pi$ stands for "product." The letters S and P, in one form or another, are used in many different contexts in mathematics.

**Exercises (6.1)**

1. Theorem. Let $A$, $B$ be finite sets, and let $a \in A$. Let $f\colon \{a\} \times B \to B$ be defined by $f((a, b)) = b$. Then $f$ is a bijection.

2. Let $A$ be the set $\{3, 5, 7\}$. List the elements of $A^2$; then list the elements of $A^3$. Find the cardinalities $|A^2|$, $|A^3|$, $|A^4|$, and $|A^5|$.

3. Find the cardinalities $|\mathscr{P}(\mathbb{N}_2 \times \mathbb{N}_3)|$ and $|\mathscr{P}(\mathbb{N}_2) \times \mathscr{P}(\mathbb{N}_3)|$.

**Remark.** Notice that for all sets $A$, $B$, $\mathscr{P}(A \times B)$ is the set of all relations from $A$ to $B$.

4. Let $n$, $k \in \mathbb{N}$, and let $A$, $B$ be sets such that $|A| = n$ and $|B| = k$. Find $|\mathscr{P}(A \times B)|$ and $|\mathscr{P}(A) \times \mathscr{P}(B)|$.

5. Theorem. Let $n \in \mathbb{N}$, and for each $1 \leq k \leq n$, let $A_k$ be a set. If for all $j$, $k \in \mathbb{N}_n$, $j \neq k$ implies $A_j \cap A_k = \varnothing$, then $\left|\bigcup_{k=1}^{n} A_k\right| = \sum_{k=1}^{n} |A_k|$.

6. Theorem. Let $n \in \mathbb{N}$ and for each $1 \leq i \leq n$, let $k_i \in \mathbb{N}$ and let $A_i$ be a set such that $|A_i| = k_i$. Then $\left|\prod_{i=1}^{n} A_i\right| = \prod_{i=1}^{n} |A_i|$.

7. Corollary. For all $n$, $k \in \mathbb{N}$, for each set $A$, if $|A| = k$ then $|A^n| = |A|^n$.

**Remark.** Up till now, we have not presented any applications of our mathematics. But combinatorics is the study of methods of counting, and so we need something to count. We can and will count such purely mathematical objects as the elements of products of finite sets. But also we will find it interesting to consider applications, or story problems, to solve by counting. (Sometimes, in fact, thinking about applications can help us to prove theorems. Applications make abstract objects seem less abstract, which helps us to visualize and analyze them.)

For the most part, we will count not concrete objects but possibilities: possible kinds of ice-cream cone, possible slates of office holders, possible outcomes of tossing a coin. Each story-problem scenario corresponds to an abstract mathematical object. Thus, we will do two sorts of counting. We will prove theorems about the number of elements in various finite sets: products of sets, sets of functions from one set to another, sets of one-to-one functions, sets of onto functions, subsets of the power set of a set. We will also solve problems about "real-world" objects. (Of course, we just make up these real-world objects.)

Now, on to the story problems.

## Applications (6.1)

1. A student has 7 mathematics books, 9 physics books, and 17 novels. The student needs to choose one from each category. There are $7 \times 9 \times 17 = 1071$ possible choices.

Abstract model. Let $M = \{m_i\}_{i=1}^{7}$, let $P = \{p_i\}_{i=1}^{9}$, and let $S = \{s_i\}_{i=1}^{17}$. Then $|M \times P \times S| = 7 \times 9 \times 17 = 1071$.

2. A coin is tossed 12 times. Each possible outcome may be represented by a sequence of 12 symbols, each a 0 or a 1, with 1 standing for heads and 0 for tails. Thus there are $2^{12}$, or 4096, distinct possible outcomes.

   Abstract model. Let $A = \{0, 1\}$. Then $|A^{12}| = 2^{12} = 4096$.

3. An ice-cream shop offers ice-cream cones with the following options: sugar cone or waffle cone; one or two scoops of ice cream; ten different flavors of ice cream: sprinkles or no sprinkles; nuts or no nuts. How many distinct sorts of ice-cream cone are offered? *[Assumptions: (1) The second scoop of ice cream is placed on top of the first. Hence a cone whose first scoop is vanilla and whose second scoop is chocolate differs from a cone with a first scoop of chocolate and a second scoop of vanilla. (2) A cone with no ice cream at all is not an ice-cream cone.]* There are 2 sorts of cone, 10 choices for the flavor of the first scoop, 11 for the second scoop (the eleventh choice is to omit the second scoop), 2 choices for sprinkles, and 2 choices for nuts. This gives $2 \times 10 \times 11 \times 2 \times 2$, or 880, distinct sorts of ice-cream cone.

   Abstract model. Let $C = \{c_i\}_{i=1}^{2}$, let $T = \{t_i\}_{i=1}^{10}$, let $S = \{s_i\}_{i=1}^{11}$, let $X = \{x_i\}_{i=1}^{2}$, and let $Y = \{y_i\}_{i=1}^{2}$. Then $|C \times T \times S \times X \times Y| = 2 \times 10 \times 11 \times 2 \times 2 = 880$.

**Notation.** Let $A$ and $B$ be sets. The symbol $F(A, B)$ denotes the set of all functions from $A$ to $B$. That is, $F(A, B) = \{f \in \mathcal{P}(A \times B) \mid f: A \to B\}$.

Exercises (6.2)

1. Theorem. Let $n, k \in \mathbb{N}$, and let $A$, $B$ be sets such that $|A| = n$ and $|B| = k$. Let $a_1, \ldots, a_n$ be the elements of $A$. Let $\theta: F(A, B) \to B^n$ be defined by $\theta(f) = (f(a_1), f(a_2), \ldots, f(a_n))$. Then $\theta$ is a bijection.

2. Corollary. Let $n, k \in \mathbb{N}$, and let $A$, $B$ be sets such that $|A| = n$ and $|B| = k$. Then there are exactly $k^n$ functions from $A$ to $B$.

3. Let $A = \{2, 3, 5, 7\}$, and let $B = \{4, 9, 11, 13, 15, 16, 20\}$. Let $f: A \to B$ be defined by $f(x) = 18 - x$. Let $a_1 = 2$, $a_2 = 3$, $a_3 = 5$, and $a_4 = 7$. Let $\theta: F(A, B) \to B^4$ be defined by $\theta(g) = (g(a_1), g(a_2), \ldots g(a_4))$. Find $\theta(f)$.

4. How many seven-digit positive integers are there?

5. A club consisting of thirty people chooses a president, secretary, and treasurer. There is no restriction on the number of offices one person can hold. How many different choices are possible? The following is an example of a choice of officers: President—Bob; Secretary—Sally; Treasurer—Sally.

6. A coin is tossed 6 times. How many distinct outcomes are possible?

7. Two dice are rolled. How many possible outcomes are there? (An outcome consists of an ordered pair $(a, b) \in \mathbb{N}_6 \times \mathbb{N}_6$, where $a$ is the value of the first die and $b$ is the value of the second.)

8. Two dice are rolled. How many outcomes are there in which the value of the first die is 4 or smaller?

**Definition and Notation.** For each nonnegative integer $k$, the symbol $k!$ is pronounced "*k factorial.*" Zero factorial equals 1; and for each nonnegative integer $k$, $(k+1)! = (k+1)(k!)$.

**Theorem (6.2)** For each $n \in \mathbb{N}$, $n! = \prod_{k=1}^{n} k$.

**Proof.** Exercise.

**Definition.** Let $A$ be a set. A *permutation* of $A$ is a bijection on $A$. We will use the symbol $S(A)$ to denote the set of permutations of $A$. We pronounce $S(A)$ as "S of A."

**Remark.** Let $n \in \mathbb{N}$, and let $A$ be a set such that $|A| = n$. Let $f \in S(A)$. Then $f$: $A \to A$ is a bijection. As Exercise 6.2.1 shows, we may label the elements of $A$ as $a_1, a_2, \ldots a_n$ and think of the function $f$ as the $n$-tuple $(f(a_1), f(a_2), \ldots f(a_n))$. Since $f$ is a bijection, this amounts to thinking of $f$ as an order in which to write the elements of $A$ in a row.

**Exercises (6.3)**

1. Theorem. For each $n \in \mathbb{N}$, $n! = \prod_{k=1}^{n} k$.

2. Theorem. For each $n \in \mathbb{N}$, $\sum_{k=1}^{n} \frac{k}{(k+1)!} = \frac{(n+1)!-1}{(n+1)!}$.

3. Theorem. For each $n \in \mathbb{N}$, if $n \geq 4$ then $2^n < n!$.

4. Theorem. For each $n \in \mathbb{N}$, if $n \geq 7$ then $3^n < n!$.

5. Let $A = \{2, 3, 5\}$. List the elements of $S(A)$.

6. Let $A = \{2, 3, 5, 7\}$. Let $B = \{f \in S(A) \mid f(5) = 5\}$. List the elements of $B$.

7. How many distinct ways are there to arrange the letters S, H, O, E in a row?

**Theorem (6.3)** Let $A$, $B$ be sets. If $|A| = |B|$ then $|S(A)| = |S(B)|$.

**Proof.** Suppose that $|A| = |B|$. Then there exists a bijection $h$: $A \to B$. We define $\theta$: $S(A) \to S(B)$ by $\theta(f) = h \circ f \circ h^{-1}$: $B \to B$.

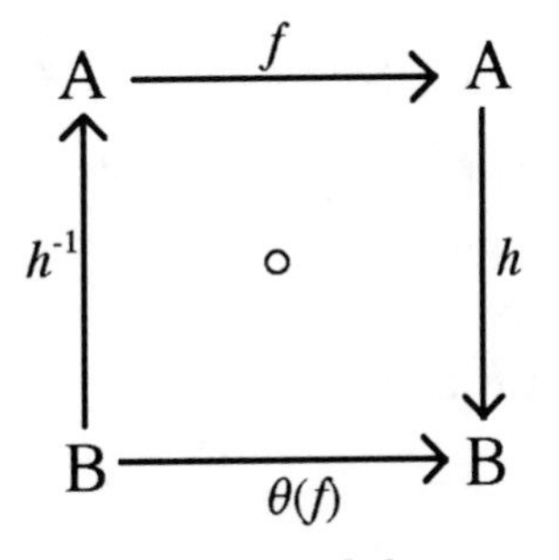


*Figure 6.2*

We first verify that for each $f \in S(A)$, $\theta(f) \in S(B)$. Let $f \in S(A)$. Then $h \circ f \circ h^{-1}$ is a function from $B$ to $B$. Since $h$, $f$, and $h^{-1}$ are all bijections, $\theta(f)$ is a bijection. Since $\theta(f)$ is a bijection on $B$, $\theta(f) \in S(B)$.

Now we will show that $\theta$ is one-to-one. Let $g_1, g_2 \in S(A)$ such that $\theta(g_1) = \theta(g_2)$. Since $\theta(g_1) = \theta(g_2)$, it follows that $h^{-1} \circ \theta(g_1) \circ h = h^{-1} \circ \theta(g_2) \circ h$. Hence $h^{-1} \circ (h \circ g_1 \circ h^{-1}) \circ h = h^{-1} \circ (h \circ g_2 \circ h^{-1}) \circ h$. That is, $g_1 = g_2$. Thus, for all $g_1, g_2 \in S(A)$, if $\theta(g_1) = \theta(g_2)$ then $g_1 = g_2$. That is, $\theta$ is one-to-one.

It remains to show that $\theta$ is onto. Let $g \in S(B)$. Then $g\colon B \to B$ is a bijection. Since $h$, $h^{-1}$, and $g$ are all bijections, $h^{-1} \circ g \circ h\colon A \to A$ is also a bijection. Let $\tilde{g} = h^{-1} \circ g \circ h$. Then $\tilde{g} \in S(A)$ and $\theta(\tilde{g}) = g$. Thus, for each $g \in S(B)$, there exists $\tilde{g} \in S(A)$ such that $\theta(\tilde{g}) = g$. Thus $\theta$ is onto.

Since $\theta$ is one-to-one and onto, $\theta$ is a bijection. Since $\theta\colon S(A) \to S(B)$ is a bijection, $|S(A)| = |S(B)|$.

Therefore, for all sets $A$, $B$, if $|A| = |B|$ then $|S(A)| = |S(B)|$. Q.E.D.

**Notation.** Let $A$ be a set, and let $x, y \in A$. We will use the symbol $\underset{y \mapsto x}{S(A)}$ to denote the set $\{f \in S(A) \mid f(y) = x\}$. The symbol $\underset{y \mapsto x}{S(A)}$ may be read as "the set of all permutations of $A$ which take $y$ to $x$."

**Theorem (6.4)** Let $A$ be a set, and let $x \in A$. Then $|\underset{x \mapsto x}{S(A)}| = |S(A - \{x\})|$.

**Proof.** Let $B = A - \{x\}$. We will show that $|\underset{x \mapsto x}{S(A)}| = |S(B)|$.

Let $f \in \underset{x \mapsto x}{S(A)}$. Since $f$ is a bijection and $f(x) = x$, $f|B$ is a bijection on $B$. Let $\theta\colon \underset{x \mapsto x}{S(A)} \to S(B)$ be defined by $\theta(f) = f|B$. We will show that $\theta$ is a bijection.

First we will show that $\theta$ is one-to-one. Let $f, g \in \underset{x \mapsto x}{S(A)}$, and suppose that $\theta(f) = \theta(g)$. Since $\theta(f) = f|B$ and $\theta(g) = g|B$, $f|B = g|B$. Since $f$,

$g \in \underset{x\mapsto x}{S(A)}$, $f(x) = x$ and $g(x) = x$. Since $f(x) = g(x)$ and $f|B = g|B$, and since $A = \{x\} \cup B$, for each $z \in A$, $f(z) = g(z)$. Since Dom $f$ = Dom $g$ $= A$ and for each $z \in A$, $f(z) = g(z)$, $f = g$. Thus for all $f, g \in \underset{x\mapsto x}{S(A)}$, if $\theta(f) = \theta(g)$ then $f = g$. That is, $\theta$ is one-to-one.

It remains to show that $\theta$ is onto. Let $h \in S(B)$. Then $h$ is a bijection on $B$. Let $g\colon A \to A$ be defined by $g(z) = \begin{cases} x, \text{ if } z = x; \\ h(z) \text{ otherwise.} \end{cases}$

Since $g|B = h$ and $h$ is a bijection on $B$, $g|B$ is a bijection on $B$. Since $g|B$ is a bijection on $B$ and $g(x) = x$, and since $A = \{x\} \cup B$, $g$ is a bijection on $A$. Since $g$ is a bijection on $A$ and $g(x) = x$, $g \in \underset{x\mapsto x}{S(A)}$. Since $g|B = h$, $\theta(g) = h$. Thus, for each $h \in S(B)$, there exists $g \in \underset{x\mapsto x}{S(A)}$ such that $\theta(g) = h$. That is, $\theta$ is onto.

Since there exists a bijection $\theta\colon \underset{x\mapsto x}{S(A)} \to S(B)$, $|\underset{x\mapsto x}{S(A)}| = |S(B)|$. Therefore, for each set $A$ and for each $x \in A$, $|\underset{x\mapsto x}{S(A)}| = |\underset{x\mapsto x}{S(A - \{x\})}|$. Q.E.D.

The following exercises will help us prove Theorem 6.5.

<u>Exercises</u> **(6.4)**

1. <u>Theorem.</u> For each set $A$, for all $x, y \in A$, $|A - \{x\}| = |A - \{y\}|$.

2. <u>Theorem.</u> For each set $A$, for all $x, y \in A$, $|\underset{x\mapsto x}{S(A)}| = |\underset{y\mapsto x}{S(A)}|$.

3. <u>Theorem.</u> Let $A$ be a set and let $x \in A$. Let $\mathscr{S} = \{\underset{y\mapsto x}{S(A)}\}_{y\in A}$. Then $\mathscr{S}$ is a partition of $S(A)$.

4. <u>Theorem.</u> For each $n \in \mathbb{N}$, for each set $A$, if there exists a partition $\mathscr{T} = \{T_k\}_{k=1}^{n}$ of $A$ such that for each $k \in \mathbb{N}_n$, $T_k$ is finite, then $|A| = \sum\limits_{k=1}^{n} |T_k|$.

5. Theorem. For each $n \in \mathbb{N}$, $|S(\mathbb{N}_n)| = \sum_{k=1}^{n} |S(\mathbb{N}_n)_{k \mapsto n}|$.

**Theorem (6.5)** For each $n \in \mathbb{N}$, $|S(\mathbb{N}_n)| = n!$.

**Proof.** The proof is by induction on $n$.

Let $n = 1$. Since $S(\mathbb{N}_1) = \{I_{\mathbb{N}_1}\}$, $|S(\mathbb{N}_1)| = 1$. Since $1! = 1$ and $|S(\mathbb{N}_1)| = 1$, $|S(\mathbb{N}_1)| = 1!$.

Let $n \in \mathbb{N}$, and suppose that $|S(\mathbb{N}_n)| = n!$. We will show that $|S(\mathbb{N}_{n+1})| = (n+1)!$. Since $\mathbb{N}_{n+1} - \{n+1\} = \mathbb{N}_n$, by Theorem 6.4 $|S(\mathbb{N}_{n+1})_{n+1 \mapsto n+1}| = |S(\mathbb{N}_n)|$. By the induction hypothesis, $|S(\mathbb{N}_n)| = n!$. Hence $|S(\mathbb{N}_{n+1})_{n+1 \mapsto n+1}| = n!$. By Exercise 2, for each $k \in \mathbb{N}_{n+1}$, $|S(\mathbb{N}_{n+1})_{k \mapsto n+1}| = |S(\mathbb{N}_{n+1})_{n+1 \mapsto n+1}|$. Thus, for each $k \in \mathbb{N}_{n+1}$, $|S(\mathbb{N}_{n+1})_{k \mapsto n+1}| = n!$. By Exercise 5, $|S(\mathbb{N}_{n+1})| = \sum_{k=1}^{n+1} |S(\mathbb{N}_{n+1})_{k \mapsto n+1}|$. Hence, $|S(\mathbb{N}_{n+1})| = \sum_{k=1}^{n+1} n! = (n+1)(n!) = (n+1)!$.

Therefore, by the principle of mathematical induction, for each $n \in \mathbb{N}$, $|S(\mathbb{N}_n)| = n!$. Q.E.D.

**Corollary.** For each $n \in \mathbb{N}$, for each set $A$, if $|A| = n$ then $|S(A)| = n!$.

**Proof.** Exercise.

**Remark.** Recall that a permutation $f: A \to A$ may be regarded as a way of writing the elements of $A$ in order in a row.

**Applications (6.2)**

1. Eight different people are to give short speeches at a graduation ceremony. There are 8!, or 40,320, possible orders in which the speakers can be scheduled.

**2.** A dance is performed in couples, each of which consists of one man and one woman. There are 7 men and 7 women in the dance hall. The number of ways to pair them off into couples for the dance is 7!, or 5040.

**3.** The letters A E I O U Y can be arranged in 6!, or 720 different orders.

**4.** Ten athletes are to sit in a row on a bench, waiting to receive awards. Six are swimmers and four are runners. All the swimmers must sit together and all the runners must sit together. How many different orders are possible, with these restrictions?

Solution: There are $2 \times 6! \times 4!$ orders possible. The factor of 2 occurs because the swimmers may occupy either the left or the right side of the bench. The number of ways to order the swimmers is 6!, and the number ways to order the runners is 4!. Since any possible overall order for the athletes involves a left/right choice, an order for the swimmers, and an order for the runners, the number of possible orders is $2 \times 6! \times 4! = 34{,}560$.

## Exercises (6.5)

**1.** Prove the Corollary to Theorem 6.5.

In exercises 2 though 5, answer each question, and show that your answer is correct.

**2.** Ten different tasks must be done at a factory, one at a time. Each task is to be done just once. How many different orders are possible for these tasks if

(a) any task may follow any other task?

(b) all of Tasks 1, 2, 3, 4, and 5 must be finished before any of tasks 6, 7, 8, 9, 10 may be begun?

(c) Task 1 must be done before Task 2 and after Task 9?

(d) Tasks 6, 7, and 8 must be done as a block, without any other tasks in between them, but may be done in any order and at any point in the proceedings?

(e) Tasks 4 and 5 must be separated by at least two other tasks?

**3.** Twenty people sit in a row for a ceremony. There are 7 married couples and 6 single people. In how many different orders is it possible for these people to sit if

(a) there are no restriction?

(b) each married couple must sit together?

(c) all the single people must sit together, but there are no restrictions on the married people?

**4.** In how many ways can we write the word SLUICE in a row

(a) without restriction?

(b) if E must either appear first or last?

(c) if the fourth letter must be C?

(d) if L and S must appear side-by-side?

(e) if L and S must not appear side-by-side?

(f) if L and S must be together, and C and U must not be together?

**5.** Ten married couples pair off for a dance. Each dance pair consists of one man and one woman. In how many ways can these people pair off to dance if

(a) there are no restrictions?

(b) one married couple, consisting of Anna and Roberto, must dance together?

(c) Anna and Roberto must not dance together?

(d) there are 6 "older" married couples and 4 "younger" married couples, and the older women must dance with the older men?

**Remark.** We recall that $\omega_0$ (omega nought) denotes the set $\mathbb{N} \cup \{0\}$.

**Definition.** Let $n \in \omega_0$, $k \in \mathbb{Z}$. If $0 \leq k \leq n$ then the number *$n$ choose $k$* (denoted by $\binom{n}{k}$ or ${}_n\mathrm{C}_k$) equals $\frac{n!}{k!(n-k)!}$. If $k < 0$ or $k > n$ then $\binom{n}{k}$ equals 0.

**Vocabulary.** The number $\binom{n}{k}$ is sometimes called the *number of combinations of $n$ objects taken $k$ at a time.*

**Theorem (6.8)** For all $n \in \omega_0$, for all $k \in \mathbb{Z}$, $\binom{n+1}{k+1} = \binom{n}{k+1} + \binom{n}{k}$.

**Proof.** Let $n \in \omega_0$, and let $k \in \mathbb{Z}$. Either $k \geq n$ or $k < 0$ or $0 \leq k < n$.

Case 1. Suppose $k \geq n$. Then $k + 1 \geq n + 1$.

Hence $\binom{n}{k+1} = 0$, and $\binom{n+1}{k+1} = \begin{cases} 1, & \text{if } k = n; \\ 0, & \text{if } k > n, \end{cases}$ and $\binom{n}{k} = \begin{cases} 1, & \text{if } k = n; \\ 0, & \text{if } k > n. \end{cases}$

Thus $\binom{n+1}{k+1} = \binom{n}{k+1} + \binom{n}{k}$.

Case 2. Suppose $k < 0$. Then $\binom{n}{k} = 0$. If $k = -1$, then $k + 1 = 0$. Hence, if $k = -1$, $\binom{n}{k+1} = \binom{n}{0} = 1$, and $\binom{n+1}{k+1} = \binom{n+1}{0} = 1$. If $k < -1$, then $\binom{n+1}{k+1} = 0$, and $\binom{n}{k+1} = 0$. Thus for all $k < 0$, $\binom{n}{k+1} = \binom{n+1}{k+1}$. It follows that, for all $k < 0$, $\binom{n+1}{k+1} = \binom{n}{k+1} + \binom{n}{k}$.

Case 3. Suppose $0 \leq k < n$. Then $\binom{n}{k} = \frac{n!}{k!(n-k)!} = \frac{n!(k+1)}{(k+1)!(n-k)!}$, and $\binom{n}{k+1} = \frac{n!}{(k+1)!(n-k-1)!} = \frac{n!(n-k)}{(k+1)!(n-k)!}$. Hence $\binom{n}{k} + \binom{n}{k+1} = \frac{n!(k+1)}{(k+1)!(n-k)!} + \frac{n!(n-k)}{(k+1)!(n-k)!} = \frac{n!(k+1+n-k)}{(k+1)!(n-k)!} = \frac{(n+1)!}{(k+1)!(n-k)!} = \binom{n+1}{k+1}$.

Therefore, for all $n \in \omega_0$, for all $k \in \mathbb{Z}$, $\binom{n+1}{k+1} = \binom{n}{k+1} + \binom{n}{k}$. Q.E.D.

**Corollary.** For all $n \in \omega_0$, for all $k \in \mathbb{Z}$, the number $\binom{n}{k}$ is an integer.

**Proof.** Exercise.

**Pascal's Triangle.** Since for all $n \in \omega_0$, for all $k \in \mathbb{Z}$, $\binom{n+1}{k+1} = \binom{n}{k+1} + \binom{n}{k}$, we can write down $\binom{n}{k}$ for all $n$, $k \in \omega_0$ in the following formation:

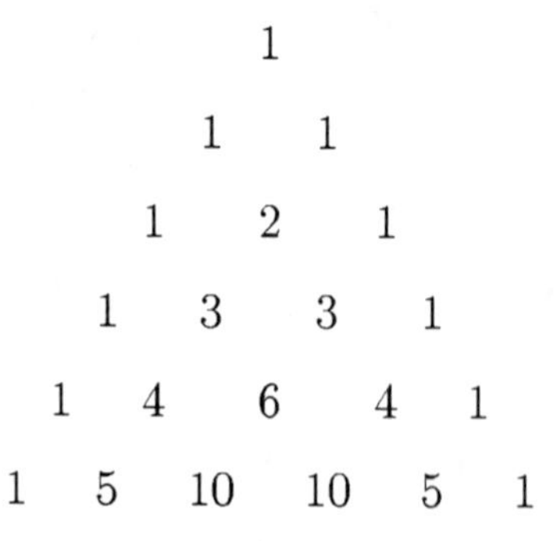

$$
\begin{array}{ccccccccccc}
 & & & & & 1 & & & & & \\
 & & & & 1 & & 1 & & & & \\
 & & & 1 & & 2 & & 1 & & & \\
 & & 1 & & 3 & & 3 & & 1 & & \\
 & 1 & & 4 & & 6 & & 4 & & 1 & \\
1 & & 5 & & 10 & & 10 & & 5 & & 1
\end{array}
$$

And so on. The 1 at the top of the triangle is $\binom{0}{0}$, and for each $n$, $k \in \omega_0$, if $k \leq n$ then the $(k+1)$st element of the $(n+1)$st row is $\binom{n}{k}$. Each number is the sum of the two numbers above it. For example, $6 = \binom{4}{2} = \binom{3}{1} + \binom{3}{2}$. So the number 6 in the figure above is the sum of the two 3's above it. Where no number is written, a zero is assumed.

This way of writing the numbers $\binom{n}{k}$ for all $n$, $k \in \omega_0$ is commonly called Pascal's triangle, although Pascal was not the first to discover it. (The triangle was discovered several times by various civilizations throughout history.)

**Notation.** Let $A$ be a set and let $k \in \mathbb{Z}$. Recall that the symbol $\mathscr{P}(A)$ denotes the power set of $A$. We will use the symbol $\mathscr{P}_k(A)$ to denote the set of all $k$-element subsets of $A$. That is, $\mathscr{P}_k(A) = \{S \in \mathscr{P}(A) \mid |S| = k\}$.

**Remark.** Since no set has negative cardinality, for each set $A$ and each negative integer $k$, $\mathscr{P}_k(A) = \varnothing$. Therefore, for all $n \in \omega_0$, for each set $A$, if $|A| = n$ then for each negative integer $k$, $|\mathscr{P}_k(A)| = \binom{n}{k} = 0$.

**Theorem (6.9)** For all $n \in \omega_0$, for all $k \in \mathbb{Z}$, for each set $A$, if $|A| = n$ then $|\mathscr{P}_k(A)| = \binom{n}{k}$.

**Proof.** The proof is by induction on $n$.

Let $A$ be a set such that $|A| = 0$. Then $A = \varnothing$, and $\mathscr{P}(A) = \mathscr{P}_0(A) = \{\varnothing\}$. Thus $|\mathscr{P}_0(A)| = 1 = \binom{0}{0}$, and for all $k \in \mathbb{Z}$, if $k \neq 0$ then $|\mathscr{P}_k(A)| = 0 = \binom{0}{k}$. Therefore, for each set $A$, if $|A| = 0$ then for each $k \in \mathbb{Z}$, $|\mathscr{P}_k(A)| = \binom{0}{k}$.

Let $n \in \omega_0$, and suppose that for each set $A$, if $|A| = n$ then for each $k \in \mathbb{Z}$, $|\mathscr{P}_k(A)| = \binom{n}{k}$. Let $B$ be a set such that $|B| = n + 1$.

Let $k \in \mathbb{N}_{n+1}$ and let $x \in B$. Let $\theta: \mathscr{P}_k(B-\{x\}) \cup \mathscr{P}_{k-1}(B-\{x\}) \to \mathscr{P}_k(B)$ be defined by $\theta(T) = \begin{cases} T, \text{ if } T \in \mathscr{P}_k(B - \{x\}); \\ T \cup \{x\}, \text{ if } T \in \mathscr{P}_{k-1}(B - \{x\}). \end{cases}$

Then $\theta$ is a bijection. (Proof of this assertion is left to the reader as an exercise).

Since there is a bijection between $\mathscr{P}_k(B - \{x\}) \cup \mathscr{P}_{k-1}(B - \{x\})$ and $\mathscr{P}_k(B)$, it follows that $|\mathscr{P}_k(B)| = |\mathscr{P}_k(B - \{x\}) \cup \mathscr{P}_{k-1}(B - \{x\})|$, which in turn equals $|\mathscr{P}_k(B - \{x\})| + |\mathscr{P}_{k-1}(B - \{x\})|$. Since $|B| = n + 1$, $|B - \{x\}| = n$. Hence, by the induction hypothesis, $|\mathscr{P}_k(B - \{x\})| = \binom{n}{k}$ and $|\mathscr{P}_{k-1}(B - \{x\})| = \binom{n}{k-1}$. Thus $|\mathscr{P}_k(B)| = \binom{n}{k} + \binom{n}{k-1} = \binom{n+1}{k}$.

We have shown that for each $1 \leq k \leq n + 1$, $\mathscr{P}_k(B) = \binom{n+1}{k}$. Since $\mathscr{P}_0(B) = \{\varnothing\}$, $|\mathscr{P}_0(B)| = 1 = \binom{n+1}{0}$. For each $k \in \mathbb{Z}$, if $k > n + 1$ or $k < 0$, then $\mathscr{P}_k(B) = \varnothing$, so $|\mathscr{P}_k(B)| = 0 = \binom{n+1}{k}$. Thus for each $k \in \mathbb{Z}$, $|\mathscr{P}_k(B)| = \binom{n+1}{k}$.

Therefore, for all $n \in \omega_0$, for all $k \in \mathbb{Z}$, for each set $A$, if $|A| = n$ then $|\mathscr{P}_k(A)| = \binom{n}{k}$. Q.E.D.

**Notation.** Let $A$ and $B$ be sets. We will use the symbol $S(A, B)$ to denote the set of one-to-one functions from $A$ to $B$. Thus, $S(A, B) = \{f \in F(A, B) \mid f \text{ is one-to-one}\}$.

**Theorem (6.10)** Let $n, k \in \mathbb{N}$, and let $A$, $B$ be sets such that $|A| = n$ and $|B| = k$. Then $|S(A, B)| = \binom{n}{k} k!$.

**Proof.** Exercise.

**Remarks.** Let $n,\ k \in \omega_0$. The number of ways to choose a $k$-element subset from a set of $n$ elements and arrange the elements of the chosen set in order is called the *number of permutations of $n$ objects taken $k$ at a time.* (Not a very catchy name.) This number is denoted by the symbol ${}_n\mathrm{P}_k$.

The number ${}_n\mathrm{P}_k = \binom{n}{k} k! = \begin{cases} 0, \text{ if } k > n; \\ \frac{n!}{(n-k)!} \text{ otherwise.} \end{cases}$

Thus, for all nonnegative integers $n$, $k$, if $n \geq k$ then ${}_n\mathrm{P}_k = \prod_{m=0}^{k-1} (n-m)$ $= n(n-1)(n-2)\ldots(n-k+1)$.

We will not use the symbol ${}_n\mathrm{P}_k$, because we prefer to think of this number as $\binom{n}{k}k!$. We mention the symbol only because it is frequently used in mathematics texts and appears as a function on many calculators.

**Applications (6.3)**

1. A group of 20 people needs to select a 3-person committee to represent the group at a meeting. The number of possible committees is $\binom{20}{3}$, or 1140.

2. A group of 20 people needs to select a president, a secretary, and a treasurer. No person may hold more than one office. The number of ways to select these three officers is $\binom{20}{3}(3!)$, or $\frac{20!}{17!}$, or 6840. There are $\binom{20}{3}$ ways to choose 3 people out of 20, and 3! ways to assign 3 different offices to 3 different people.

3. A coin is tossed 10 times. There are $2^{10}$, or 1024, possible outcomes, each of which may be represented by a 10-character sequence $\{a_k\}_{k=1}^{10}$

   where $a_k = \begin{cases} 1, \text{ if the } k\text{th toss is heads;} \\ 0, \text{ otherwise.} \end{cases}$

The number of possible outcomes with exactly 5 heads is the same as the number of 5-element subsets of $\mathbb{N}_{10}$. Thus the number of outcomes with exactly 5 heads is $\binom{10}{5}$, or 252.

4. The number of ways to arrange the letters ALLUVIAL is $\binom{8}{3}\binom{5}{2}3!$, or $\frac{8!}{3!2!}$, or 3360. There are $\binom{8}{3}$ ways to choose the positions of the L's, $\binom{5}{2}$ ways to position the $A$'s once the L's are in place, and 3! arrangements for the remaining 3 letters.

**Notation.** Let $B$ be a set and let $n \in \mathbb{N}$. Then $\mathcal{P}_n(B) = \{C \in \mathcal{P}(B) \mid |C| = n\}$. For each $C \in \mathcal{P}_n(B)$, $S(C) = \{f: C \to C \mid f \text{ is a bijection}\}$. Thus the symbol $\bigcup_{C \in \mathcal{P}_n(B)} S(C)$ denotes the set $\{f: C \to C \mid f \text{ is a bijection and } C \in \mathcal{P}_n(B)\}$. This notation occurs in Exercises 3 and 4 below.

**Exercises (6.6)**

1. Theorem. For all $n \in \omega_0$, for all $k \in \mathbb{Z}$, $\binom{n}{k}$ is an integer.

2. Theorem. Let $n \in \omega_0$, and let $B$ be a set such that $|B| = n + 1$. Let $k \in \mathbb{N}_{n+1}$, and let $\theta: \mathcal{P}_k(B - \{x\}) \cup \mathcal{P}_{k-1}(B - \{x\}) \to \mathcal{P}_k(B)$ be defined by
$$\theta(T) = \begin{cases} T, \text{ if } T \in \mathcal{P}_k(B - \{x\}); \\ T \cup \{x\}, \text{ if } T \in \mathcal{P}_{k-1}(B - \{x\}). \end{cases}$$
Then $\theta$ is a bijection.

3. Theorem. Let $n, k \in \mathbb{N}$, and let $B$ be a set such that $|B| = k$. Then $\left| \bigcup_{C \in \mathcal{P}_n(B)} S(C) \right| = \binom{n}{k} k!$.

4. Theorem. Let $n, k \in \mathbb{N}$, and let $A$ and $B$ be sets such that $|A| = n$ and $|B| = k$. Since $|A| = n$, there exists a bijection $g: \mathbb{N}_n \to A$. For each $C \in \mathcal{P}_n(B)$, since $|C| = n$, there exists a bijection $h_C: C \to \mathbb{N}_n$. Let $\theta: S(A, B) \to \bigcup_{C \in \mathcal{P}_n(B)} S(C)$ be defined by $\theta(f) = f \circ g \circ h_{f(A)}$. Then

$\theta$ is a bijection. [Hint: First show that for each $f \in S(A,\, B)$, $\theta(f) \in S(f(A))$. Thus, for each $f \in S(A,\, B)$, $\theta(f) \in \bigcup_{\substack{C \in \\ \mathscr{P}_n(B)}} S(C)$. Then prove that $\theta$ is a bijection. It may help to draw a diagram with arrows to represent functions.]

$$f(A) \xrightarrow{h_{f(A)}} \mathbb{N}_n \xrightarrow{g} A \xrightarrow{f} f(A)$$

**5.** Prove Theorem 6.10.

**6.** How many possible arrangements are there for the letters HIPPOPOTAMUS?

**7.** A coin is tossed 7 times. How many possible outcomes are there with 5 or more heads?

**8.** How many arrangements are there of the letters HIPPOPOTAMUS in which no two P's are side by side?

**9.** A coin is tossed 7 times. In how many of the possible outcomes are there exactly 2 heads in the first three tosses and exactly 2 heads in the last four tosses?

**10.** <u>Theorem</u>. For all $n \in \mathbb{N}$, $\sum_{k=0}^{n} \binom{n}{k} = 2^n$.

We will use the results of Exercises 11 through 16 in our upcoming proof of the binomial theorem. Many of these problems are very simple, once you understand the notation.

**11.** <u>Theorem</u>. For all $a,\, b,\, c \in \mathbb{Z}$, for each $f\colon \mathbb{Z} \to \mathbb{R}$, $\sum_{k=a+c}^{b+c} f(k) = \sum_{k=a}^{b} f(k+c)$. [Hint: Use induction on $b-a$, for $b-a \in \omega_0$.]

**12.** <u>Theorem</u>. For each $n \in \mathbb{N}$, for all $x$, $y \in \mathbb{R}$, $\sum_{k=0}^{n} \binom{n}{k} x^{k+1} y^{n-k} + \sum_{k=0}^{n} \binom{n}{k} x^k y^{n+1-k} = \sum_{k=0}^{n-1} \binom{n}{k} x^{k+1} y^{n-k} + \sum_{k=n}^{n} \binom{n}{k} x^{k+1} y^{n-k} + \sum_{k=0}^{0} \binom{n}{k} x^k y^{n+1-k} + \sum_{k=1}^{n} \binom{n}{k} x^k y^{n+1-k}$.

**13.** <u>Theorem</u>. For all $n \in \mathbb{N}$, for all $x$, $y \in \mathbb{R}$, $\sum_{k=n}^{n} \binom{n}{k} x^{k+1} y^{n-k} = \sum_{k=n+1}^{n+1} \binom{n+1}{k} x^k y^{n+1-k}$.

**14.** <u>Theorem</u>. For each $n \in \mathbb{N}$, for all $x$, $y \in \mathbb{R}$, $\sum_{k=0}^{0} \binom{n}{k} x^k y^{n+1-k} = \sum_{k=0}^{0} \binom{n+1}{k} x^k y^{n+1-k}$.

**15.** <u>Theorem</u>. For each $n \in \mathbb{N}$, for all $x$, $y \in \mathbb{R}$, $\sum_{k=0}^{n-1} \binom{n}{k} x^{k+1} y^{n-k} = \sum_{k=1}^{n} \binom{n}{k-1} x^k y^{n+1-k}$.

**16.** <u>Theorem</u>. For each $n \in \mathbb{N}$, for all $x$, $y \in \mathbb{R}$, $\sum_{k=0}^{n-1} \binom{n}{k} x^{k+1} y^{n-k} + \sum_{k=1}^{n} \binom{n}{k} x^k y^{n+1-k} = \sum_{k=1}^{n} \binom{n+1}{k} x^k y^{n+1-k}$.

**17.** <u>Theorem</u>. For each $n \in \mathbb{N}$, for all $x$, $y \in \mathbb{R}$, $\sum_{k=0}^{n} \binom{n}{k} x^{k+1} y^{n-k} + \sum_{k=0}^{n} \binom{n}{k} x^k y^{n+1-k} = \sum_{k=0}^{n+1} \binom{n+1}{k} x^k y^{n+1-k}$.

We are now ready to prove a very interesting and useful result, the binomial theorem.

Given two real numbers, $x$, $y$, the binomial theorem allows us to calculate, say, $(x+y)^7$, without having to multiply $(x+y)$ by itself several times. The binomial theorem implies that $(x+y)^7 = \sum_{k=0}^{7} \binom{7}{k} x^k y^{7-k} = y^7 + 7xy^6 + 21x^2y^5 + 35x^3y^4 + 35x^4y^3 + 21x^5y^2 + 7x^6y + x^7$.

Without further ado, we proceed to state and prove the theorem.

**Theorem (6.11)** (Binomial theorem) For each $n \in \mathbb{N}$, for all $x$, $y \in \mathbb{R}$, $(x+y)^n = \sum_{k=0}^{n} \binom{n}{k} x^k y^{n-k}$.

**Proof.** The proof is by induction on $n$.

Let $x$, $y \in \mathbb{R}$. Since $\sum_{k=0}^{1} \binom{1}{k} x^k y^{1-k} = y + x$ and $(x+y)^1 = x+y = y+x$, $(x+y)^1 = \sum_{k=0}^{1} \binom{1}{k} x^k y^{1-k}$.

Let $n \in \mathbb{N}$, and suppose that for all $x$, $y \in \mathbb{R}$, $(x+y)^n = \sum_{k=0}^{n} \binom{n}{k} x^k y^{n-k}$. Let $x$, $y \in \mathbb{R}$. Then $(x+y)^{n+1} = (x+y)\,(x+y)^n = (x+y)\sum_{k=0}^{n} \binom{n}{k} x^k y^{n-k} = x\sum_{k=0}^{n} \binom{n}{k} x^k y^{n-k} + y\sum_{k=0}^{n} \binom{n}{k} x^k y^{n-k} = \sum_{k=0}^{n} \binom{n}{k} x^{k+1} y^{n-k} + \sum_{k=0}^{n} \binom{n}{k} x^k y^{n+1-k}$. By Exercise 17, $\sum_{k=0}^{n} \binom{n}{k} x^{k+1} y^{n-k} + \sum_{k=0}^{n} \binom{n}{k} x^k y^{n+1-k} = \sum_{k=0}^{n+1} \binom{n+1}{k} x^k y^{n+1-k}$. Thus $(x+y)^{n+1} = \sum_{k=0}^{n+1} \binom{n+1}{k} x^k y^{n+1-k}$.

Therefore, by the principle of mathematical induction, for all $n \in \mathbb{N}$, for all $x$, $y \in \mathbb{R}$, $(x+y)^n = \sum_{k=0}^{n} \binom{n}{k} x^k y^{n-k}$. Q.E.D.

**Vocabulary.** Because of the binomial theorem, the numbers $\binom{n}{k}$ are sometimes called *binomial coefficients.*

**Exercises (6.7).**

1. Let $x, y \in \mathbb{R}$. Use the binomial theorem to expand the product $(x-1)^9$.

2. Let $x, y \in \mathbb{R}$. Use the binomial theorem to expand the product $(2x^3 - 3yx^2)^9$.

3. The binomial theorem implies that for each $n \in \mathbb{N}$, $11^n = (10+1)^n = \sum_{k=0}^{n} \binom{n}{k} 10^k$. Thus $11^1 = 11$, $11^2 = 121$, $11^3 = 1331$, and so on. The powers of 11 can be read off from the rows of Pascal's triangle. For each $n \in \mathbb{N}_7$, use Pascal's triangle to find $11^n$.

4. Theorem. For each $n \in \mathbb{N}$, $\sum_{k=0}^{n} \binom{n}{k} = 2^n$. [We have proved this before, but now we can prove it in a new way, by using the binomial theorem to expand $(1+1)^n$.]

5. Theorem. For each $n \in \mathbb{N}$, $\sum_{k=0}^{n} (-1)^k \binom{n}{k} = 0$.

6. Theorem. For each $n, k \in \mathbb{N}$, $k\binom{n}{k} = n\binom{n-1}{k-1}$.

7. Theorem. For each $n \in \mathbb{N}$, $\sum_{k=1}^{n} k\binom{n}{k} = n2^{n-1}$.

8. Theorem. For each $n \in \mathbb{N}$, $\sum_{k=1}^{n} (-1)^k k\binom{n}{k} = 0$.

9. Theorem. For each $n \in \mathbb{N}$, let $F_n$ denote the $n$th Fibonacci number. Then for each $n \in \mathbb{N}$, $\sum_{k=0}^{n} \binom{n-k}{k} = F_{n+1}$.

10. Theorem. Let $m, n \in \mathbb{N}$, and let $A$, $B$ be sets such that $|A| = m$, $|B| = n$, and $A \cap B = \varnothing$. Let $r \in \mathbb{N}_{m+n}$, and let $\psi$: $\mathscr{P}_r(A \cup B) \to \bigcup_{k=0}^{r} (\mathscr{P}_k(A) \times \mathscr{P}_{r-k}(B))$ be defined by $\psi(T) = (T \cap A, T \cap B)$. Then $\psi$ is a bijection.

11. Theorem. For all $m, n \in \mathbb{N}$, for all $r \in \mathbb{N}_{m+n}$, $\binom{m+n}{r} = \sum_{k=0}^{r} \binom{m}{k}\binom{n}{r-k}$.

12. Theorem. For each $n \in \mathbb{N}$, $\binom{2n}{n} = \sum_{m=0}^{n} \binom{n}{m}^2$.

**13.** Let $n,\ k \in \mathbb{N}$ such that $n \geq 2$ and $1 \leq k \leq n-1$. Then the number $\binom{n}{k}$ is surrounded by six other numbers in Pascal's triangle, thus:

$$\begin{array}{ccccc} & \binom{n-1}{k-1} & & \binom{n-1}{k} & \\ \binom{n}{k-1} & & \binom{n}{k} & & \binom{n}{k+1} \\ & \binom{n+1}{k} & & \binom{n+1}{k+1} & \end{array}$$

If we consider these six numbers to be the vertices of a hexagon, and number the vertices clockwise from any one of the six, we notice that the product of the odd-numbered vertices equals the product of the even-numbered vertices. Prove this result.

Theorem. For each $n,\ k \in \mathbb{N}$ such that $n \geq 2$ and $1 \leq k \leq n-1$,
$\binom{n-1}{k-1}\binom{n}{k+1}\binom{n+1}{k} = \binom{n-1}{k}\binom{n+1}{k+1}\binom{n}{k-1}$.

**Notation.** Sums and products of real numbers may be indexed by any finite set. Let $n \in \mathbb{N}$, and let $A$ be a set such that $|A| = n$. Since $|A| = n$, the elements of $A$ may be indexed by $\mathbb{N}_n$, so that $A = \{x_k\}_{k=1}^{n}$. Let $f\colon A \to \mathbb{R}$. The notation $\sum\limits_{x \in A} f(x)$ means the same thing as $\sum\limits_{k=1}^{n} f(x_k)$. Similarly, $\prod\limits_{x \in A} f(x) = \prod\limits_{k=1}^{n} f(x_k)$.

Let $n \in \mathbb{N}$, and for each $k \in \mathbb{N}_n$, let $A_k$ be a set, and let $x_k,\ y_k \in \mathbb{R}$. For each $S \in \mathscr{P}(\mathbb{N}_n)$, we will use the symbol $\prod x_S$ as an abbreviation for $\prod\limits_{k \in S} x_k$. Similarly, we will use $\sum x_S$ for $\sum\limits_{k \in S} x_k$, $\bigcap A_S$ for $\bigcap\limits_{k \in S} A_k$, and $\bigcup A_S$ for $\bigcup\limits_{k \in S} A_k$.

Summation and product signs are often used in tandem. For example, let $x_1, x_2, y_1, y_2 \in \mathbb{R}$. Then

$$\sum\left(\prod x_S \prod y_{\mathbb{N}_2 - S}\right)_{\mathscr{P}(\mathbb{N}_2)} = \sum_{S \in \mathscr{P}(\mathbb{N}_2)} \left(\prod x_S \prod y_{\mathbb{N}_2 - S}\right) = \sum_{S \in \mathscr{P}(\mathbb{N}_2)} \left(\prod_{k \in S} x_k \prod_{r \in \mathbb{N}_2 - S} y_r\right)$$

$$= \left(\prod_{k \in \varnothing} x_k \prod_{r \in \mathbb{N}_2} y_r\right) + \left(\prod_{k \in \{1\}} x_k \prod_{r \in \{2\}} y_r\right) + \left(\prod_{k \in \{2\}} x_k \prod_{r \in \{1\}} y_r\right) + \left(\prod_{k \in \mathbb{N}_2} x_k \prod_{r \in \varnothing} y_r\right)$$

$$= y_1 y_2 + x_1 y_2 + x_2 y_1 + x_1 x_2.$$

In the next set of exercises, we will use this notation to prove the binomial theorem in a new way. In Chapter 7, this notation will help us to prove the inclusion-exclusion principle.

## Exercises (6.8)

1. Theorem. For each $k \in \mathbb{N}_3$, for all $x_k, y_k \in \mathbb{R}$,

$$\sum\left(\prod x_S \prod y_{\mathbb{N}_3 - S}\right)_{\mathscr{P}(\mathbb{N}_3)}$$

$$= y_1 y_2 y_3 + x_1 y_2 y_3 + y_1 x_2 y_3 + y_1 y_2 x_3$$
$$+ x_1 x_2 y_3 + x_1 y_2 x_3 + y_1 x_2 x_3 + x_1 x_2 x_3.$$

2. Theorem. For all $n \in \mathbb{N}$, $\mathscr{P}(\mathbb{N}_n) \cup (\mathscr{P}(\mathbb{N}_{n+1}) - \mathscr{P}(\mathbb{N}_n)) = \mathscr{P}(\mathbb{N}_{n+1})$, and $\mathscr{P}(\mathbb{N}_n) \cap (\mathscr{P}(\mathbb{N}_{n+1}) - \mathscr{P}(\mathbb{N}_n)) = \varnothing$.

3. Theorem. For all $n \in \mathbb{N}$, there exists a bijection $f\colon \mathbb{N}_{2^{n+1}} \to \mathscr{P}(\mathbb{N}_{n+1})$ such that $f(\mathbb{N}_{2^n}) = \mathscr{P}(\mathbb{N}_n)$ and $f(\mathbb{N}_{2^{n+1}} - \mathbb{N}_{2^n}) = \mathscr{P}(\mathbb{N}_{n+1}) - \mathscr{P}(\mathbb{N}_n)$.

4. Theorem. Let $n \in \mathbb{N}$, and for each $k \in \mathbb{N}_{n+1}$, let $x_k, y_k \in \mathbb{R}$. Then

$$\sum\left(\prod x_S \prod y_{\mathbb{N}_{n+1} - S}\right)_{\mathscr{P}(\mathbb{N}_n)} + \sum\left(\prod x_S \prod y_{\mathbb{N}_{n+1} - S}\right)_{(\mathscr{P}(\mathbb{N}_{n+1}) - \mathscr{P}(\mathbb{N}_n))}$$

$$= \sum\left(\prod x_S \prod y_{\mathbb{N}_{n+1} - S}\right)_{\mathscr{P}(\mathbb{N}_{n+1})}.$$

**5.** Theorem. Let $n \in \mathbb{N}$, and for each $k \in \mathbb{N}_{n+1}$, let $x_k, y_k \in \mathbb{R}$. Then

$$x_{n+1} \sum(\prod x_S \prod y_{\mathbb{N}_n - S})_{\mathscr{P}(\mathbb{N}_n)}$$
$$= \sum(\prod x_S \prod y_{\mathbb{N}_{n+1} - S})_{(\mathscr{P}(\mathbb{N}_{n+1}) - \mathscr{P}(\mathbb{N}_n))}.$$

**6.** Theorem. Let $n \in \mathbb{N}$, and for each $k \in \mathbb{N}_{n+1}$, let $x_k, y_k \in \mathbb{R}$. Then

$$y_{n+1} \sum(\prod x_S \prod y_{\mathbb{N}_n - S})_{\mathscr{P}(\mathbb{N}_n)}$$
$$= \sum(\prod x_S \prod y_{\mathbb{N}_{n+1} - S})_{\mathscr{P}(\mathbb{N}_n)}.$$

**7.** Theorem. Let $n \in \mathbb{N}$, and for each $k \in \mathbb{N}_{n+1}$, let $x_k, y_k \in \mathbb{R}$. Then

$$(x_{n+1} + y_{n+1}) \sum(\prod x_S \prod y_{\mathbb{N}_n - S})_{\mathscr{P}(\mathbb{N}_n)}$$
$$= \sum(\prod x_S \prod y_{\mathbb{N}_{n+1} - S})_{\mathscr{P}(\mathbb{N}_{n+1})}.$$

**8.** Theorem. Let $n \in \mathbb{N}$, and for each $k \in \mathbb{N}_n$, let $x_k, y_k \in \mathbb{R}$. Then

$$\prod_{k=1}^{n} (x_k + y_k) = \sum(\prod x_S \prod y_{\mathbb{N}_n - S})_{\mathscr{P}(\mathbb{N}_n)}.$$

**9.** Theorem. Let $n \in \mathbb{N}$, let $x, y \in \mathbb{R}$, and for each $k \in \mathbb{N}_n$, let $x_k = x$, $y_k = y$. Then for each $m \in \mathbb{N}_n \cup \{0\}$,

$$\sum(\prod x_S \prod y_{\mathbb{N}_n - S})_{\mathscr{P}_m(\mathbb{N}_n)} = \binom{n}{m} x^m y^{n-m}.$$

**10.** Theorem. (Binomial theorem) For all $n \in \mathbb{N}$, for all $x, y \in \mathbb{R}$,

$$(x + y)^n = \sum_{m=0}^{n} \binom{n}{m} x^m y^{n-m}.$$

**11.** Theorem. Let $n \in \mathbb{N}$, and for each $k \in \mathbb{N}_n$, let $x_k, y_k \in \mathbb{R}$. For each $m \in \mathbb{N}_{2^n}$, for each $k \in \mathbb{N}_n$, let

$$a_{m,k} = \begin{cases} x_k, & \text{if there exists } j \in \mathbb{N}_{2^{k-1}} \text{ such that } m \equiv j \bmod 2^k; \\ y_k & \text{otherwise.} \end{cases}$$

Then $\prod_{k=1}^{n} (x_k + y_k) = \sum_{m=1}^{2^n} (\prod_{k=1}^{n} a_{m,k})$.

# Chapter 7

# Derangements and Other Entertainments

*Improbabilities are apt to be overestimated. It is true that I should have been surprised in the past to learn that Professor Hardy had joined the Oxford Group. But one could not say the adverse chance was $10^6$ : 1. Mathematics is a dangerous profession; an appreciable proportion of us go mad, and then this particular event would be quite likely.*

*J. E. Littlewood,* Littlewood's Miscellany

*Print is one extreme of typographical development, the other being mathematical notation. It consists, in the occident anyway, of the representation of sounds by purely arbitrary shapes, and arranging them so that those in the know can reproduce the spoken words intended. This process is known as Reading, and is very uncommon in adults. It is uncommon because, firstly, it is in many cases frankly impossible, the number of phonetical symbols being inadequate; secondly, because of the extreme familiarity of the word-shapes to a population whose experience is necessarily derived in the main from marks printed on paper.*

*It is this second circumstance, familiarity with the word or phrase shapes, that has led to the unpremeditated birth of a visual language.*

*Now you (yes,* you*) before you tear this paper into little bits, kindly tell me whether that last paragraph was written by me as part of my satanic campaign against decency and reason or whether it is taken from a book written in all seriousness by some other person. On your answer to that query will depend more than I would care to say in public.*

*Myles na gGopaleen (Flann O'Brien),* The Best of Myles

*"Algebra, like laudanum, deadens pain," Fritz wrote. "But the study of algebra has confirmed for me that philosophy and mathematics, like mathematics and music, speak the same language. That, of course, is not enough. I shall see my way in time. Patience, the key will turn."*

*Penelope Fitzgerald,* The Blue Flower

The *inclusion-exclusion principle* is entertaining to use but difficult to introduce. The difficulty lies in the notation. Before stating this principle in full generality, we will consider a few special cases.

**Theorem (7.1)** Let $A$ and $B$ be finite sets. Then $|A \cup B| = |A| + |B| - |A \cap B|$.

**Proof.** Since $A \cup B = A \cup (B - A)$, and since $A \cap (B - A) = \varnothing$, $|A \cup B| = |A| + |B - A|$. Since $B = (A \cap B) \cup (B - A)$ and since $(A \cap B) \cap (B - A) = \varnothing$, $|B| = |A \cap B| + |B - A|$. Thus $|B - A| = |B| - |B \cap A|$. Therefore $|A| + |B| - |B \cap A| = |A| + |B - A| = |A \cup B|$. Q.E.D.

**Remark.** Here's another way to think of it. When we add $|A|+|B|$, for each $x \in A \cap B$, $x$ gets counted twice. So we need to subtract $|A \cap B|$ from $|A| + |B|$ to get $|A \cup B|$.

**Theorem (7.2)** Let $A$, $B$, and $C$ be finite sets. Then $|A \cup B \cup C| = |A| + |B| + |C| - (|A \cap B| + |B \cap C| + |A \cap C|) + |A \cap B \cap C|$.

**Proof.** Let A, B and C be finite sets. Since B and C are finite, $B \cup C$ is finite. Hence, by Theorem 7.1, $|A \cup B \cup C| = |A \cup (B \cup C)| = |A| + |B \cup C| - |A \cap (B \cup C)|$. Again by Theorem 7.1, $|B \cup C| = |B| + |C| - |B \cap C|$. Thus $|A \cup B \cup C| = |A| + |B| + |C| - |B \cap C| - |A \cap (B \cup C)|$.

Now $A \cap (B \cup C) = (A \cap B) \cup (A \cap C)$, so $|A \cap (B \cup C)| = |(A \cap B) \cup (A \cap C)|$. By Theorem 7.1, $|(A \cap B) \cup (A \cap C)| = |A \cap B| + |A \cap C| - |(A \cap B) \cap (A \cap C)|$.

Since $(A \cap B) \cap (A \cap C) = A \cap B \cap C$, $|A \cap (B \cup C)| = |A \cap B| + |A \cap C| - |A \cap B \cap C|$.

Therefore, $|A \cup B \cup C| = |A| + |B| + |C| - |B \cap C| - (|A \cap B| + |A \cap C| - |A \cap B \cap C|) = |A| + |B| + |C| - (|A \cap B| + |B \cap C| + |A \cap C|) + |A \cap B \cap C|$. Q.E.D.

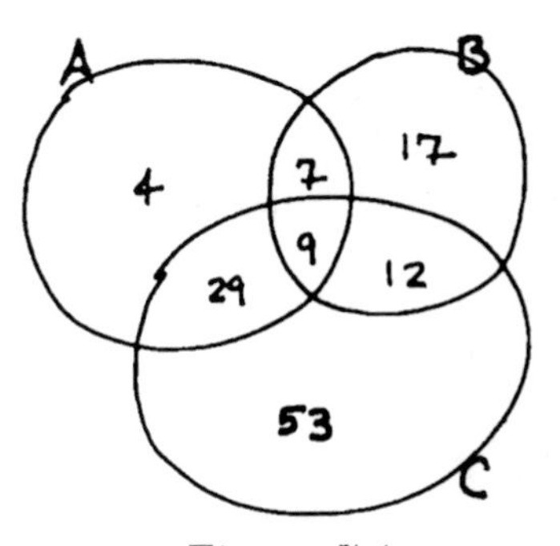


*Figure 7.1*

**Remark.** When we add $|A| + |B| + |C|$, the elements of the sets $A \cap B$, $A \cap C$, and $B \cap C$ get counted twice. So we subtract $|A \cap B| + |B \cap C| + |A \cap C|$. But now the elements of $A \cap B \cap C$ have been added 3 times and subtracted 3 times. So $A \cap B \cap C$ has not been counted, and we need to add $|A \cap B \cap C|$. Thus we have $|A \cup B \cup C| = |A| + |B| + |C| - (|A \cap B| + |B \cap C| + |A \cap C|) + |A \cap B \cap C|$.

**Exercises (7.1)**

1. The diagram on the previous page shows sets $A$, $B$, $C$ such that $|A - (B \cup C)| = 4$, $|(A \cap B) - C| = 7$, $|B - (A \cup C)| = 17$, $|(A \cap C) - B| = 29$, $|A \cap B \cap C| = 9$, $|(B \cap C) - A| = 12$, and $|C - (A \cup B)| = 53$. Thus $|A \cup B \cup C| = 4 + 7 + 17 + 29 + 9 + 12 + 53 = 131$. Find the numbers $|A|$, $|B|$, $|C|$, $|A \cap B|$, $|B \cap C|$, $|A \cap C|$, and $|A \cap B \cap C|$, and recalculate $|A \cup B \cup C|$ using Theorem 7.2.

2. Theorem. Let $A_1$, $A_2$, $A_3$, and $A_4$ be finite sets. Then $|\bigcup_{k=1}^{4} A_k| = \sum_{k=1}^{4} |A_k| - (|A_1 \cap A_2| + |A_1 \cap A_3| + |A_1 \cap A_4| + |A_2 \cap A_3| + |A_2 \cap A_4| + |A_3 \cap A_4|) + (|A_1 \cap A_2 \cap A_3| + |A_1 \cap A_2 \cap A_4| + |A_1 \cap A_3 \cap A_4| + |A_2 \cap A_3 \cap A_4|) - |A_1 \cap A_2 \cap A_3 \cap A_4|$.

3. Explain the theorem of Exercise 2 by imitating the Remarks following Theorem 7.1 and Theorem 7.2.

**Notation.** Let $A_1$, $A_2$, $A_3$, $A_4$ be finite sets. Using the notation of the last section of Chapter 6,

$$|A_1 \cap A_2| + |A_1 \cap A_3| + |A_1 \cap A_4| + |A_2 \cap A_3| + |A_2 \cap A_4| + |A_3 \cap A_4|$$

may be written as

$$\sum |\bigcap A_S|_{\mathcal{P}_2(\mathbb{N}_4)}.$$

Recall that $\mathbb{N}_4 = \{1, 2, 3, 4\}$ and that $\mathcal{P}_2(\mathbb{N}_4) = \{S \subseteq \mathbb{N}_4 \mid |S| = 2\}$. Thus $\mathcal{P}_2(\mathbb{N}_4) = \{\{1, 2\}, \{1, 3\}, \{1, 4\}, \{2, 3\}, \{2, 4\}, \{3, 4\}\}$.

Hence

$$\sum |\bigcap A_S|_{\wp_2(\mathbb{N}_4)}$$
$$= |A_1 \cap A_2| + |A_1 \cap A_3| + |A_1 \cap A_4|$$
$$+ |A_2 \cap A_3| + |A_2 \cap A_4| + |A_3 \cap A_4|.$$

Exercise 2 shows that for all finite sets $A_1$, $A_2$, $A_3$, $A_4$,

$$|\bigcup_{k=1}^{4} A_k|$$
$$= |A_1| + |A_2| + |A_3| + |A_4|$$
$$- (|A_1 \cap A_2| + |A_1 \cap A_3| + |A_1 \cap A_4|$$
$$+ |A_2 \cap A_3| + |A_2 \cap A_4| + |A_3 \cap A_4|)$$
$$+ (|A_1 \cap A_2 \cap A_3| + |A_1 \cap A_2 \cap A_4|$$
$$+ |A_1 \cap A_3 \cap A_4| + |A_2 \cap A_3 \cap A_4|)$$
$$- |A_1 \cap A_2 \cap A_3 \cap A_4|.$$

That is,

$$|\bigcup_{k=1}^{4} A_k|$$
$$= \sum |\bigcap A_S|_{\wp_1(\mathbb{N}_4)} - \sum |\bigcap A_S|_{\wp_2(\mathbb{N}_4)}$$
$$+ \sum |\bigcap A_S|_{\wp_3(\mathbb{N}_4)} - \sum |\bigcap A_S|_{\wp_4(\mathbb{N}_4)}.$$

Since for each $j \in \mathbb{N}$, $(-1)^{j+1} = 1$ if $j$ is odd and $(-1)^{j+1} = -1$ if $j$ is even, we may abbreviate the statement of the theorem as follows: For all finite sets $A_1$, $A_2$, $A_3$, $A_4$,

$$|\bigcup_{k=1}^{4} A_k| = \sum_{j=1}^{4} (-1)^{j+1} \sum |\bigcap A_S|_{\wp_j(\mathbb{N}_4)}.$$

**Remark.** The inclusion-exclusion principle, which we will prove as Theorem 7.3, may be stated as follows.

**Theorem.** Let $n \in \mathbb{N}$, and for each $k \in \mathbb{N}_n$, let $A_k$ be a finite set. Then

$$|\bigcup_{k=1}^{n} A_k| = \sum_{j=1}^{n}(-1)^{j+1}\sum|\bigcap A_S|_{\wp_j(\mathbb{N}_n)}.$$

The proof of this theorem is by induction, and is quite straightforward. The only difficulty lies in using the notation: double summation signs, intersection and union signs, cardinality signs, and so on. In the next set of exercises, we invite the reader to do almost all the work of the proof. Experience has shown that it is very hard to understand the notation unless you are the one who is actually writing and using it. Exercises 7.2 will be used essentially in our proof of the inclusion-exclusion principle.

**Exercises (7.2)**

1. Theorem. Let $A_1$ be a finite set. Then $(\bigcap A_S)_{\wp_1(\mathbb{N}_1)} = A_1$, and

$$\sum_{j=1}^{1}(-1)^{j+1}\sum|\bigcap A_S|_{\wp_1(\mathbb{N}_1)} = |A_1|.$$

2. Theorem. For each $j \in \mathbb{N}$, $(-1)^{j+1} = \begin{cases} 1, & \text{if } j \text{ is odd;} \\ -1 & \text{otherwise.} \end{cases}$

3. Theorem. Let $n \in \mathbb{N}$, and for each $k \in \mathbb{N}_{n+1}$, let $A_k$ be a finite set. Then $|\bigcup_{k=1}^{n+1} A_k| = |\bigcup_{k=1}^{n} A_k| + |A_{n+1}| - |\bigcup_{k=1}^{n}(A_k \cap A_{n+1})|$.

4. Theorem. Let $n \in \mathbb{N}$, and for each $k \in \mathbb{N}_{n+1}$, let $A_k$ be a finite set. Then

$$\sum|\bigcap A_S|_{\wp_1(\mathbb{N}_n)} = \sum_{k=1}^{n}|A_k|,$$

$$\text{and } \sum|\bigcap A_S|_{\wp_1(\mathbb{N}_{n+1})} = \sum_{k=1}^{n}|A_k| + |A_{n+1}|.$$

5. Theorem. Let $n \in \mathbb{N}$ and let $j \in \mathbb{N}_n$. Then

$$\mathscr{P}_{j+1}(\mathbb{N}_{n+1}) - \mathscr{P}_{j+1}(\mathbb{N}_n) = \{S \cup \{n+1\} \mid S \in \mathscr{P}_j(\mathbb{N}_n)\}.$$

6. Theorem. Let $n \in \mathbb{N}$, and for each $k \in \mathbb{N}_{n+1}$, let $A_k$ be a finite set. Then for each $j \in \mathbb{N}_n$,

$$\sum |\bigcap (A_S \cap A_{n+1})|_{\mathscr{P}_j(\mathbb{N}_n)}$$

$$= \sum |\bigcap A_{S \cup \{n+1\}}|_{\mathscr{P}_j(\mathbb{N}_n)}$$

$$= \sum |\bigcap A_S|_{(\mathscr{P}_{j+1}(\mathbb{N}_{n+1}) - \mathscr{P}_{j+1}(\mathbb{N}_n))}.$$

7. Theorem. Let $n \in \mathbb{N}$, and for each $k \in \mathbb{N}_{n+1}$, let $A_k$ be a finite set. Then

$$\sum_{j=2}^{n} (-1)^{j+1} \sum |\bigcap A_S|_{\mathscr{P}_j(\mathbb{N}_n)} - \sum_{j=1}^{n} (-1)^{j+1} \sum |\bigcap A_S|_{(\mathscr{P}_{j+1}(\mathbb{N}_{n+1}) - \mathscr{P}_{j+1}(\mathbb{N}_n))}$$

$$= \sum_{j=2}^{n+1} (-1)^{j+1} \sum |\bigcap A_S|_{\mathscr{P}_j(\mathbb{N}_{n+1})}.$$

8. Theorem. Let $n \in \mathbb{N}$, and for each $k \in \mathbb{N}_{n+1}$, let $A_k$ be a finite set.

Then

$$\sum_{j=2}^{n} (-1)^{j+1} \sum |\bigcap A_S|_{\mathscr{P}_j(\mathbb{N}_n)}$$

$$- \sum_{j=1}^{n} (-1)^{j+1} \sum |\bigcap (A_S \cup A_{n+1})|_{\mathscr{P}_j(\mathbb{N}_n)}$$

$$= \sum_{j=2}^{n+1} (-1)^{j+1} \sum |\bigcap A_S|_{\mathscr{P}_j(\mathbb{N}_{n+1})}.$$

9. Theorem. Let $n \in \mathbb{N}$, and for each $k \in \mathbb{N}_{n+1}$, let $A_k$ be a finite set.

Then

$$\sum_{j=1}^{n} (-1)^{j+1} \sum |\bigcap A_S|_{\mathscr{P}_j(\mathbb{N}_n)} + |A_{n+1}|$$

$$- \sum_{j=1}^{n}(-1)^{j+1}\sum|\bigcap(A_S \cap A_{n+1})|_{\mathscr{P}_j(\mathbb{N}_n)}$$

$$= \sum_{j=1}^{n+1}(-1)^{j+1}\sum|\bigcap A_S|_{\mathscr{P}_j(\mathbb{N}_{n+1})}.$$

**10.** Theorem. Let $n \in \mathbb{N}$, and for each $k \in \mathbb{N}_{n+1}$, let $A_k$ be a finite set. If

$$|\bigcup_{k=1}^{n} A_k| = \sum_{j=1}^{n}(-1)^{j+1}\sum|\bigcap A_S|_{\mathscr{P}_j(\mathbb{N}_n)},$$

then

$$|\bigcup_{k=1}^{n+1} A_k| = \sum_{j=1}^{n+1}(-1)^{j+1}\sum|\bigcap A_S|_{\mathscr{P}_j(\mathbb{N}_{n+1})}.$$

Now that you have done all the work, we are ready to prove the inclusion-exclusion principle.

**Theorem (7.3)** (Inclusion-exclusion principle.) Let $n \in \mathbb{N}$, and for each $k \in \mathbb{N}_n$, let $A_k$ be a finite set. Then $|\bigcup_{k=1}^{n} A_k| = \sum_{j=1}^{n}(-1)^{j+1}\sum|\bigcap A_S|_{\mathscr{P}_j(\mathbb{N}_n)}$.

**Proof.** The proof is by induction on $n$.

Let $n = 1$, and let $A_1$ be a finite set. Since $\bigcup_{k=1}^{1} A_1 = A_1$, $|\bigcup_{k=1}^{1} A_1| = |A_1|$. By Exercise 1,

$$\sum_{j=1}^{1}(-1)^{j+1}\sum|\bigcap A_S|_{\mathscr{P}_1(\mathbb{N}_1)} = |A_1|.$$

Hence,

$$|\bigcup_{k=1}^{1} A_k| = \sum_{j=1}^{1}(-1)^{j+1}\sum|\bigcap A_S|_{\mathscr{P}_1(\mathbb{N}_1)}.$$

Let $n = 2$. Let $A_1$ and $A_2$ be finite sets. By Theorem 7.1,

$$|\bigcup_{k=1}^{2} A_k| = |A_1| + |A_2| - |A_1 \cap A_2| = \sum_{j=1}^{2}(-1)^{j+1}\sum|\bigcap A_S|_{\mathcal{P}_j(\mathbb{N}_2)}.$$

Let $n \in \mathbb{N}$, $n \geq 2$, and suppose that for all finite sets $B_k$, $1 \leq k \leq n$, $|\bigcup_{k=1}^{n} B_k| = \sum_{j=1}^{n}(-1)^{j+1}\sum|\bigcap A_S|_{\mathcal{P}_j(\mathbb{N}_n)}$. For each $k \in \mathbb{N}_{n+1}$, let $A_k$ be a finite set. By the induction hypothesis, $|\bigcup_{k=1}^{n} A_k| = \sum_{j=1}^{n}(-1)^{j+1}\sum|\bigcap A_S|_{\mathcal{P}_j(\mathbb{N}_n)}$. Hence, by Exercise 10, $|\bigcup_{k=1}^{n+1} A_k| = \sum_{j=1}^{n+1}(-1)^{j+1}\sum|\bigcap A_S|_{\mathcal{P}_j(\mathbb{N}_{n+1})}$.

Therefore, by the principle of mathematical induction, for all $n \in \mathbb{N}$, for each collection of finite sets $\{A_k\}_{k=1}^{n}$,

$$|\bigcup_{k=1}^{n} A_k| = \sum_{j=1}^{n}(-1)^{j+1}\sum|\bigcap A_S|_{\mathcal{P}_j(\mathbb{N}_n)}.$$

Q.E.D.

Remark. In many elementary textbooks, the inclusion-exclusion principle is stated as follows.

$$\text{For each } n \in \mathbb{N},\ |\bigcup_{k=1}^{n} A_k| = \sum_{j=1}^{n}(-1)^{j+1}\sum_{\substack{1 \leq k_1 < k_2 < \\ \dots < k_j \leq n}}\cdots\sum|\bigcap_{i=1}^{j} A_{k_i}|.$$

We have used different notation for two reasons. First, the usual notation is even more cumbersome to read and write than the notation we have used. Secondly, we are already accustomed to the set $\mathbb{N}_n$ and the concept of power sets. The more mathematics we know, and the more concepts we understand, the more convenient we can make our notation.

## Exercises (7.3)

1. Theorem. Let $n \in \mathbb{N}$ and let $j \in \mathbb{N}_n$. Then $\mathcal{P}_j(\mathbb{N}_n) = \{\{k_i\}_{i=1}^{j} \mid$ for all $i \in \mathbb{N}_j$, $k_i \in \mathbb{N}$ and $1 \leq k_1 < k_2 \ldots < k_j \leq n\}$.

2. Theorem. Let $n \in \mathbb{N}$ and for each $k \in \mathbb{N}_n$, let $A_k$ be a finite set. Let $x \in \bigcup_{k=1}^{n} A_k$ and let $f\colon \mathcal{P}(\mathbb{N}_n) \to \{0, 1\}$ be defined by

$$f(S) = \begin{cases} 1 \text{ if } x \in \bigcap A_S; \\ 0 \text{ otherwise.} \end{cases}$$

Then for each $x \in \bigcup_{k=1}^{n} A_k$, $\sum((-1)^{|S|+1} f(S))_{\mathcal{P}(\mathbb{N}_n)} = 1$.

[Hint: For each $x \in \bigcup_{k=1}^{n} A_k$ there exists $m \in \mathbb{N}_n$ such that $|\{S \in \mathcal{P}(\mathbb{N}_n) \mid x \in S\}| = m$. It follows that for each $j \in \mathbb{N}_n$,

$$|\{S \in \mathcal{P}_j(\mathbb{N}_n) \mid x \in \bigcap A_S\}| \;=\; \binom{m}{j}.$$

Remark. Exercise 2 shows that in the adding and subtracting of the inclusion-exclusion principle, for each $x \in \bigcup_{k=1}^{n} A_k$, the element $x$ winds up getting counted exactly once. To see this, we write

$$\sum((-1)^{|S|+1} f(S))_{\mathcal{P}(\mathbb{N}_n)} \;=\; \sum_{j=0}^{n}(-1)^{j+1} \sum f(S)_{\mathcal{P}_j(\mathbb{N}_n)}$$

$$= \sum_{j=1}^{n}(-1)^{j+1} \sum f(S)_{\mathcal{P}_j(\mathbb{N}_n)} \;=\; \sum_{j=1}^{n}((-1)^{j+1} \sum_{k \in S} 1)_{\{S \in \mathcal{P}_j(\mathbb{N}_n) \mid x \in S\}}.$$

**Counting onto functions.** Let $A$ and $B$ be sets. We recall that $F(A, B)$ is the set of all functions from $A$ to $B$, and $S(A, B)$ is the set of all one-to-one functions from $A$ to $B$. We will use $T(A, B)$ to denote the set of all onto functions from $A$ to $B$.

Let $n, k \in \mathbb{N}$ and $A$, $B$ be sets such that $|A| = n$ and $|B| = k$. We have shown that $|F(A, B)| = k^n$ and $|S(A, B)| = \binom{k}{n} n!$. How many onto functions are there from $A$ to $B$? If $|A| < |B|$, then $|T(A, B)| = 0$. If $|A| = |B|$, then $|T(A, B)| = n!$

We invite the reader to prove the following theorem.

**Theorem (7.4)** Let $n, k \in \mathbb{N}$ and let $A$, $B$ be sets such that $|A| = n$ and $|B| = k$. Then the number of onto functions from A to B is $\sum_{j=1}^{n} \binom{k}{j}(-1)^{k-j} j^n$.

**Proof.** Exercise.

Exercises 7.4 will guide you through one proof of Theorem 7.4. But if you are feeling adventurous, feel free to do Exercise 8 without doing Exercises 1 through 7 first.

**Exercises (7.4)**

1. Theorem. Let $A$ and $B$ be sets. Then $T(A, B) = F(A, B) - \bigcup_{x \in B} F(A, B - \{x\})$.

2. Theorem. Let $A$ and $B$ be sets, and let $S \in B$. Then

$$\bigcap_{x \in S} F(A, B - \{x\}) = F(A, (B - S)).$$

3. Theorem. Let $n$, $k$, $j \in \mathbb{N}$, such that $n \geq k > j$. Let $A$, $B$, $S$ be sets such that $S \subseteq B$ and $|A| = n$, $|B| = k$, and $|S| = j$. Then $|F(A, (B - S))| = (k - j)^n$.

4. Theorem. Let $n$, $k \in \mathbb{N}$, such that $n \geq k$, and let $A$, $B$ be sets such that $|A| = n$, $|B| = k$. Then $|\bigcup_{x \in B} F(A, B - \{x\})| = \sum_{j=1}^{k-1} (-1)^{j+1} \sum |F(A, B - S)|_{\wp_j(B)}$.

5. Theorem. Let $n, k \in \mathbb{N}$, such that $n \geq k$, and let $A$, $B$ be sets such that $|A| = n$, $|B| = k$. Then for each $j \in \mathbb{N}_{k-1}$,

$$\sum |F(A, B - S)|_{\wp_j(B)} = \binom{k}{j}(k - j)^n.$$

**6.** Theorem. Let $n, k \in \mathbb{N}$, such that $n \geq k$, and let $A, B$ be sets such that $|A| = n$, $|B| = k$. Then $|T(A, B)| = \sum_{j=0}^{k-1}(-1)^j\binom{k}{j}(k-j)^n$.

**7.** Theorem. For all $n, k \in \mathbb{N}$, $\sum_{j=0}^{k-1}(-1)^j\binom{k}{j}(k-j)^n = \sum_{j=1}^{k}(-1)^{k-j}\binom{k}{j}j^n$.

**8.** Theorem. Let $n, k \in \mathbb{N}$, such that $n \geq k$, and let $A, B$ be sets such that $|A| = n$, $|B| = k$. Then $|T(A, B)| = \sum_{j=1}^{k}(-1)^{k-j}\binom{k}{j}j^n$.

**9.** For each $1 \leq k \leq 3$, find the number of onto functions from $\mathbb{N}_3$ to $\mathbb{N}_k$.

**10.** For each $1 \leq k \leq 4$, find the number of onto functions from $\mathbb{N}_4$ to $\mathbb{N}_k$.

**11.** For each $1 \leq k \leq 5$, find the number of onto functions from $\mathbb{N}_5$ to $\mathbb{N}_k$.

**12.** Theorem. Let $n \in \mathbb{N}$, and let $A$ be a set such that $|A| = n$. For each $k \in \mathbb{N}_n$, let $F_k(A) = \{f\colon A \to A \mid |f(A)| = k\}$. Then for each $k \in \mathbb{N}_n$, $|F_k(A)| = \binom{n}{k}\sum_{j=1}^{k}(-1)^{k-j}\binom{k}{j}j^n$.

**13.** Four people compete for prizes in 7 different contests. These 4 people are the only contestants. Each person competes in every contest. One prize is awarded in each contest. There are no ties.

(**a**) How many possible outcomes are there for the awarding of prizes? (An outcome consists of the numbers 1 through 7, with each number followed by the name of the person who wins the contest with that number.)

(**b**) In how many of these possible outcomes does each contestant win at least one prize?

(**c**) In how many possible outcomes do exactly three people win prizes?

(d) In how many possible outcomes do exactly 2 people win prizes?

(e) In how many outcomes is there only one prize winner?

(f) Add the results of Exercises (b) through (e). Do you get the result of Exercise (a)?

Notation. We recall that for each set $A$, $S(A)$ is the set of permutations of $A$, and that for each $x \in A$, $\underset{x \mapsto x}{S(A)} = \{f \in S(A) \mid f(x) = x\}$.

Definition. Let $n \in \mathbb{N}$, and let $A$ be a set such that $|A| = n$. The set of *derangements* of $A$, denoted by $D(A)$, is the set $\{f \in S(A) \mid$ for each $x \in A$, $f(x) \neq x\}$. For each $n \in \mathbb{N}$, the symbol $D(n)$ denotes the number of derangements of a set of $n$ elements.

Example (7.1) Each of 8 people writes his/her name on a piece of paper and drops it into a hat. The hat is shaken, and then each person draws out one of the name-bearing pieces of paper. The number of possible outcomes is 8!. The number of possible outcomes in which no one draws his/her own name is $D(8)$, the number of derangements of a set of 8 elements.

Exercises (7.5)

1. Theorem. For each finite set $A$, $D(A) = S(A) - \bigcup_{x \in A} \underset{x \mapsto x}{S(A)}$.

2. Theorem. Let $n, k \in \mathbb{N}$ such that $k \leq n$, and let $A, B$ be sets such that $B \subseteq A$ and $|A| = n$ and $|B| = k$. Then $|\bigcap_{x \in B} \underset{x \mapsto x}{S(A)}| = (n-k)!$

3. Theorem. Let $n \in \mathbb{N}$ and let $A$ be a set such that $|A| = n$. Then $|D(A)| = |S(A)| - \sum_{k=1}^{n} (-1)^{k+1} \sum | \bigcap_{x \in B} \underset{x \mapsto x}{S(A)} |_{\mathscr{P}_k(A)}$.

4. Theorem. Let $n \in \mathbb{N}$. Then $D(n) = \sum_{k=0}^{n} \binom{n}{k} (-1)^k (n-k)!$

$= \sum_{k=0}^{n} \binom{n}{k} (-1)^{n-k} k!$.

5. Theorem. Let $n \in \mathbb{N}$. Then $D(n) = n! \sum_{k=2}^{n} (-1)^k \frac{1}{k!}$.

6. Find $D(n)$ for all $1 \leq n \leq 7$.

7. Theorem. For all $n \in \mathbb{N}$, $n! = 1 + \sum_{k=1}^{n} \binom{n}{k} D(k)$.

8. Twelve people, consisting of six married couples, pair off for a dance. Each dance pair consists of one man and one woman.

   (a) How many ways are there for these 6 men and 6 women to pair off for the dance?

   (b) How many ways are there for these people to pair off for the dance, if no married couple dances together?

   (c) For each $1 \leq k \leq 6$, how many ways are there for these people to pair off for the dance, if exactly $k$ married couples dance together?

9. Twelve people, consisting of six married couples, sit in a row. For each $1 \leq k \leq 6$, let $C_k$ denote the $k$th married couple.

   (a) How many ways are there for these 12 people to sit in a row?

   (b) How many ways are there for these people to sit in a row if couple $C_1$ sits side by side?

   (c) How many ways are there for these people to sit in a row if couple $C_1$ sits together and couple $C_2$ sits together?

   (d) Let $1 \leq j \leq 6$ and let $S \in \mathcal{P}_j(\mathbb{N}_n)$. How many ways are there for these people to sit in a row if for each $i \in S$ couple $C_i$ sits together?

   (e) How many ways are there for these 12 people to sit in a row if no married couple sits together?

**Combinatorics in probability.** Counting is a very fundamental idea in the theory of probability. Here we will offer a very brief introduction to the use of combinatorics in probability.

Probability concerns *outcomes* of *experiments*. These two notions are basic and undefined. Loosely speaking, an *experiment* is a situation that can turn out in more than one way. An *outcome* is one of the ways in which an experiment can turn out.

For example, tossing a coin once is an experiment. This experiment has two possible outcomes: heads, tails. Tossing a coin five times in succession is another experiment. The set of possible outcomes of this experiment is the set of all possible sequences of five symbols, each of which is either $H$ (for heads) or $T$ (for tails). Thus, there are 32 distinct outcomes for this experiment.

**Definition.** A *sample space* is a set of all possible outcomes of an experiment.

**Definition.** An *event* is a subset of a sample space.

**Example (7.2)** An experiment consists of tossing a coin three times in succession. Let $C = \{H, T\}$ and let $S = C^3$. Then $S$ is a sample space for this experiment.

Let $A$ be the event that the experiment results in exactly 2 heads and exactly 1 tail.

Since $S = \{HHH, HHT, HTH, HTT, THH, THT, TTH, TTT\}$, the event $A = \{HTH, HHT, THH\}$.

**Choice of sample space.** For the experiment of tossing a coin 3 times, we could use the sample space $T = \{0, 1, 2, 3\}$. Here 0 represents the outcome that no heads result, 1 that exactly one head results, and so on. The event $A$, that exactly 2 heads result, is the set $\{2\}$.

Intuitively, the set $S$ seems to be a more informative sample space than the set $T$. The set $S$ has more elements, and hence more subsets, than $T$. Hence the sample space $S$ allows us to describe more events than the sample space $T$.

Another big advantage of $S$ over $T$ as a sample space is that the elements of $S$ are *equiprobable*, if the coin we toss is fair. A *fair coin* is a coin which, when tossed, is equally likely to come up heads or tails. *Equiprobable* means equally likely.

**Definition.** Let $S$ be a finite sample space for an experiment, such that the elements of $S$ are equiprobable. Let $A$ be a subset of $S$. Then the *probability* of the event $A$ is the number $\frac{|A|}{|S|}$. The symbol $P(A)$ denotes the probability of the event $A$.

**Examples (7.3)**

1. An experiment consists of tossing a fair coin three times. Let $C = \{H, T\}$. The sample space $S = C^3$ contains eight equiprobable outcomes. For each $i \in \mathbb{N}_3$, let $A_i$ be the event that exactly $i$ heads occur in 3 tosses. Then $P(A_0) = \frac{1}{8}$, $P(A_1) = \frac{3}{8}$, $P(A_2) = \frac{3}{8}$, and $P(A_3) = \frac{1}{8}$.

2. An experiment consists of rolling a pair of fair dice once. (Each die is a cube; the six sides are labeled with the numbers 1 through 6; since the dice are fair, each die is as likely to land showing one number as another.) We use the sample space $S = \mathbb{N}_6 \times \mathbb{N}_6$, because the outcomes of $S$ are equiprobable. We think of one of the dice as the first die, and the other as the second die. An outcome is an ordered pair whose first coordinate is the number shown on the first die and whose second coordinate is the number shown on the second die.

*Figure 7.2:* the outcome $(2, 4)$

Since $S = \mathbb{N}_6 \times \mathbb{N}_6$, $|S| = 6 \times 6 = 36$.

Let $A$ be the event that the sum of the values shown by the dice is 7. Then $A = \{(1,6), (2,5), (3,4), (4,3), (5,2), (6,1)\}$, so $|A| = 6$. Thus $P(A) = \frac{6}{36} = \frac{1}{6}$.

## Exercises (7.6)

1. A pair of fair dice is rolled once. For each $2 \leq k \leq 12$, let $A_k$ be the event that the sum of the values shown on the dice is $k$.

   (a) For each $2 \leq k \leq 12$, find $P(A_k)$.

   (b) Find $\sum_{k=2}^{12} P(A_k)$.

2. A pair of fair dice is rolled once. Let $E$ be the event that the sum of the values is at most 10, and let $F$ be the event that the sum of the values is at least 5.

   (a) Find $P(E)$.
   (b) Find $P(F)$.
   (c) Find $P(E \cap F)$.
   (d) Find $P(E - F)$.
   (e) Find $P(F - E)$.
   (f) Find $P(E \cup F)$.

3. A fair coin is tossed 4 times. For each $0 \leq k \leq 4$, let $A_k$ be the probability that exactly $k$ heads result.

   (a) For each $0 \leq k \leq 4$, find $P(A_k)$.

   (b) Find $\sum_{k=0}^{4} P(A_k)$.

4. Theorem. Let $n, k \in \mathbb{N}$. Suppose that a fair coin is tossed $n$ times. Then the probability that exactly $k$ tosses result in heads is $\frac{\binom{n}{k}}{2^n}$.

5. A fair coin is tossed 7 times. For each $0 \leq k \leq 7$, let $A_k$ be the event that exactly $k$ heads result.

   (a) For each $0 \leq k \leq 7$, find $P(A_k)$.

   (b) Find $\sum_{k=0}^{7} P(A_k)$.

   (c) For each $0 \leq k \leq 7$, let $B_k$ be the event that *at least* $k$ heads come up in 7 tosses. For each $0 \leq k \leq 7$, find $P(B_k)$.

6. <u>Theorem.</u> Let $S$ be a finite sample space for an experiment, such that all outcomes in $S$ are equiprobable. Let $A, B \subseteq S$. Then $P(A \cup B) = P(A) + P(B) - P(A \cap B)$.

7. <u>Theorem.</u> Let $S$ be a finite sample space for an experiment, such that all outcomes in $S$ are equiprobable. Let $n \in \mathbb{N}$, and for each $k \in \mathbb{N}_n$, let $A_k \subseteq S$. Then $P(\bigcup_{k=1}^{n} A_k) = \sum_{j=1}^{n} (-1)^{j+1} \sum P(\bigcap A_S)_{\mathscr{P}_j(\mathbb{N}_n)}$.

8. <u>Theorem.</u> Let $S$ be a finite sample space for an experiment, and suppose that all outcomes in $S$ are equiprobable. Let $A \subseteq S$ and let $A^C = S - A$. Then $P(A^C) = 1 - P(A)$.

9. Each of eight people writes his/her name on a separate slip of paper. Then the slips of paper are folded, tossed into a hat and shaken. Each person draws a slip of paper at random.

   (a) Find the probability that no one draws his or her own name.

   (b) For each $0 \leq k \leq 8$, let $A_k$ be the event that exactly $k$ people draw their own names. Find $P(A_k)$ for each $k \in \mathbb{N}_8$.

   (c) Find $\sum_{k=0}^{8} P(A_k)$.

10. Seven married couples (i.e., fourteen people) sit in a row in random order.

    (a) Find the probability that no married couple sits side by side.

(**b**) For each $k \in \mathbb{N}_7$, let $A_k$ be the event that exactly $k$ married couples sit side by side. For each $k \in \mathbb{N}_7$, find $P(A_k)$.

(**c**) Find $\sum_{k=0}^{7} P(A_k)$.

**11.** Theorem. Let $S$ be a finite sample space for a probability experiment, such that the outcomes of $S$ are equiprobable. Let $n \in \mathbb{N}$, and for each $k \in \mathbb{N}_n$, let $A_k$ be a set such that $\{A_k\}_{k=1}^{n}$ is a partition of $S$. Then $\sum_{k=1}^{n} \mathscr{P}(A_k) = 1$.

**12.** A club consisting of 35 people chooses a committee of 4 at random. This club consists of 19 women and 16 men.

(**a**) Let $C$ be the set of club members, and let $S_1$ be the sample space whose elements are all the 4-element subsets of $C$. That is, $S_1 = \mathscr{P}_4(C)$. What is $|S_1|$?

(**b**) Let $A_1 \subseteq S_1$ be the set of 4-element subsets of $C$ which contain 2 men and 2 women. What is $|A_1|$?

(**c**) Find $P(A_1)$.

(**d**) Let $S_2$ be the sample space $\{(x_1, x_2, x_3, x_4) \in C^4 \mid$ for all $i, j \in \mathbb{N}_4$, if $i \neq j$ then $x_i \neq x_j\}$. Find $|S_2|$.

(**e**) Let $A_2 = \{(x_1, x_2, x_3, x_4) \in S_2 \mid$ exactly two of the $x_i$'s are women and exactly two are men$\}$. Find $|A_2|$.

(**f**) Find $P(A_2)$.

**Remarks.** Exercise 12 illustrates the fact that there may be more than one possible sample space for an experiment. The probability that a committee of 4 chosen randomly from a group of 16 men and 19 women consists of two members of each gender does not depend on the sample space used.

Probability is a large and fascinating subject. Only the merest taste of probability is offered in this chapter. We refer the interested reader to an introductory undergraduate course in probability.

Probability uses calculus in a very interesting way. To really appreciate probability, it's good to know calculus well. If you are a person who needs to prove theorems before doing computations, perhaps you should study real analysis first, then calculus, then probability.

The pigeonhole principle. We conclude our introduction to combinatorics with a brief discussion of the pigeonhole principle. We have already mentioned this principle in connection with one-one and onto functions. The pigeonhole principle (that if there are more pigeons than pigeonholes, then each pigeon cannot have a hole all to itself) is a counting principle from the realm of common sense. It sometimes plays a surprising role in mathematical problems.

**Example (7.4)** An old riddle which has appeared in both joke books and mathematical texts runs as follows.

There are 11 yellow socks and 17 red socks in a drawer. A person reaches into the drawer in the dark and pulls out some socks at random. How many socks must the person take to be sure of getting at least 2 socks of the same color?

The answer, of course, is 3. There are only two colors, red and yellow; hence, given any 3 socks, at least 2 must be the same color. (The numbers 11 and 17 have nothing to do with the solution of the problem.)

**Example (7.5)** Let $S$ be a set of 51 distinct positive integers less than or equal to 99. Then there exist $x$, $y \in S$ such that $x \neq y$ and $x + y = 100$.

Let $T = \{100 - x \mid x \in S\}$. Let $A = S - \{50\}$ and let $B = T - \{50\}$. Then $|A| \geq 50$ and $|B| \geq 50$ and $|A \cup B| \leq 99$. Therefore $A \cap B \neq \varnothing$. Thus there exists $x \in A \cap B$. Since $x \in B$, there exists $y \in S$ such that $x = 100 - y$. Since $x \in B$, $x \neq 50$. Thus $x \neq y$. Thus, there exist $x$, $y \in S$ such that $x \neq y$ and $x + y = 100$.

## Exercises (7.7)

1. There are 117 socks in a large bin. There are 29 blue socks, 11 red socks, 57 green socks, 18 yellow socks, and 2 white socks. What is the smallest number of socks that someone reaching into the drawer in the dark must grab to be sure of getting at least 2 of the same color?

2. Theorem. Let $n \in \mathbb{N}$, and let $A \subseteq \mathbb{N}$ such that $|A| > n$. Then there exist $x$, $y \in A$ such that $x \equiv y \bmod n$.

3. Theorem. Let $n \in \mathbb{N}$ and let $A$ be a subset of $\mathbb{N}$ such that $|A| = n+1$ and for each $x \in A$, $x \leq 2n$. Then there exist $m$, $k \in A$ such that $m \neq k$ and $m + k = 2n$.

4. A rails-to-trails society is building a hiking trail 100 miles long. There is a marker at the beginning of the trail and a milestone at every mile from the beginning. The society plants 150 trees along the trail, at least one within each marked mile. Prove that there exist $i$, $j \in \mathbb{N}_{100}$ such that exactly 43 trees are planted between the $i$th and $j$th milestones. (No tree is planted at a milestone.)

**Decimal expansions of rational numbers.** The pigeonhole principle explains why decimal expansion of rational numbers are periodic, while those of irrational numbers are not.

The following theorem describes a process that is already familiar to us: division of positive integers, with remainders. The theorem was proved by Euclid.

**Theorem (7.5)** Let $n \in \omega_0$, $m \in \mathbb{N}$. Then there exist unique $q$, $r \in \omega_0$ such that $n = mq + r$ and $r < m$.

**Proof.** Let $S = \{n - mk \mid k \in \omega_0\}$. Since $n \in S \cap \omega_0$, the set $S \cap \omega_0$ is nonempty. Let $r$ be the least element of $S \cap \omega_0$. Then, by definition of $S$, there exists $q \in \omega_0$ such that $n = mq + r$.

It remains to show that $r < m$. By way of contradiction, suppose that $r \geq m$. Then there exists $t \in \omega_0$ such that $r = m + t$.

Since $n = qm + r$, it follows that $n = qm + m + t$, and thus that $n = (q + 1)m + t$. Since $q \in \omega_0$, also $q + 1 \in \omega_0$. Since $q + 1 \in \omega_0$, it follows that $n - (q + 1)m \in S$. Thus $t \in S$. Since $t \in S$ and $t \in \omega_0$, $t \in S \cap \omega_0$. Since $m \in \mathbb{N}$ and $r = m + t$, it follows that $r > t$. But since $t \in S \cap \omega_0$ and $r$ is the least element of $S \cap \omega_0$, $r \leq t$. Thus $r > t$ and $r \leq t$. $\rightarrow\leftarrow$

Thus, for all $n \in \omega_0$, $m \in \mathbb{N}$, there exist $q, r \in \omega_0$ such that $n = qm + r$ and $r < m$. The uniqueness of the numbers $q$ and $r$ is left to the reader as an exercise. Therefore, for all $n \in \omega_0$, $m \in \mathbb{N}$, there exist unique $q$, $r \in \omega_0$ such that $n = qm + r$ and $r < m$. Q.E.D.

**Exercises (7.8)**

1. Finish the proof of Theorem 7.5 by proving the following.

   Theorem. Let $n \in \omega_0$, $m \in \mathbb{N}$. We have shown that there exist $q$, $r \in \omega_0$ such that $n = qm + r$ and $r < m$. The numbers $q$ and $r$ are unique.

Exercises 2 – 6 use Theorem 7.5 and the pigeonhole principle to prove a surprising result.

2. Theorem. For each $n$, $m \in \mathbb{N}$, there exists a unique $r \in \omega_0$ such that $n = r \bmod m$ and $r < m$.

3. Theorem. Let $n$, $m \in \mathbb{N}$. If $m$ is not a perfect square then there exist $x_1$, $y_1, x_2$, $y_2 \in \omega_0$ such that:

(**a**) $x_1, y_1, x_2, y_2 < \sqrt{m}$;

(**b**) $x_1 n - y_1 \equiv x_2 n - y_2 \bmod m$;

(**c**) $x_1 \neq x_2$ or $y_1 \neq y_2$.

[Hint: Let $S = \{x \in \omega_0 \mid x < \sqrt{m}\}$. Since $0 \in S$ and $m$ is not a perfect square, $|S| > \sqrt{m}$, so $|S^2| > m$. Apply the pigeonhole principle to the set $S^2$.]

4. Theorem. Let $n, m \in \mathbb{N}$. If $m$ is not a perfect square then there exist $x, y \in \mathbb{Z}$, not both zero, such that $nx \equiv y \bmod m$ and $|x|$, $|y| < \sqrt{m}$.

5. Theorem. Let $n, m \in \mathbb{N}$. If $m$ is not a perfect square and $m \mid n^2 + 1$ then there exist $x, y \in \mathbb{Z}$, not both zero, such that $-x^2 \equiv y^2 \bmod m$ and $|x|, |y| < \sqrt{m}$.

6. Theorem. Let $m, n \in \mathbb{N}$ such that $m$ is not a perfect square and $m \mid n^2 + 1$. Then there exist $a, b \in \mathbb{N}$ such that $a^2 + b^2 = m$.

**Example (7.6)** Since $13 \mid 5^2 + 1$, there exist $x, y \in \mathbb{N}$ such that $x^2 + y^2 = 13$. Let $x = 2$ and let $y = 3$; then $x^2 + y^2 = 13$.

**Remark.** The theorem of Exercise 4 is called Thue's Lemma, and was proved by Axel Thue. The theorem of Exercise 6 was proved by Fermat.

We now apply the result of Theorem 7.5 to decimal expansions of real numbers.

**Remark.** As we did in Chapter 5, we use the symbols $x = \sum_{k=1}^{\infty} 10^{-k} a_k$, where for each $k \in \mathbb{N}$, $a_k \in \omega_0$ and $a_k \leq 9$, to mean that the number $x$ has a decimal expansion in which the $k$th digit after the decimal point is $a_k$. We do not consider infinite sums in general. We assume naively, as we did in our youth, that a decimal point followed by an infinite string of digits denotes a real number.

When we add and subtract decimal expansions, we restrict ourselves to cases in which no "carrying" or "borrowing" is necessary. When we multiply an expansion by a power of ten, we move the decimal point. These are the only operations that our naive intuition allows us to perform.

**Theorem (7.6)** Let $x$ be a real number such that $0 < x < 1$. For each $k \in \mathbb{N}$, let $0 \leq a_k \leq 9$ such that $\sum_{k=1}^{\infty} 10^{-k} a_k$ is a decimal expansion for $x$. If there exist $j \in \omega_0$, $p \in \mathbb{N}$ such that, for all $k \in \mathbb{N}$, $k > j$ implies $a_k = a_{k+p}$, then $x \in \mathbb{Q}$.

In fact, $x = \dfrac{\sum_{k=1}^{j+p} 10^{j+p-k} a_k - \sum_{k=1}^{j} 10^{j-k} a_k}{10^j(10^p - 1)}$.

**Proof.** The next set of exercises will help the reader to prove this theorem.

**Exercises (7.9)**

1. Theorem. For each $k \in \mathbb{N}$, let $a_k \in \omega_0$ such that $a_k \leq 9$. Let $x = \sum_{k=1}^{\infty} 10^{-k} a_k$. Let $j \in \omega_0$ and let $p \in \mathbb{N}$. Then $10^{j+p} x = \sum_{k=1}^{j+p} 10^{j+p-k} a_k + \sum_{k=j+p+1}^{\infty} 10^{j+p-k} a_k$, and $10^j x = \sum_{k=1}^{j} 10^{j-k} a_k + \sum_{k=j+1}^{\infty} 10^{j-k} a_k$.

2. Theorem. For each $k \in \mathbb{N}$, let $a_k \in \omega_0$ such that $a_k \leq 9$. If there exist $j \in \omega_0$, $p \in \mathbb{N}$ such that for each $k \in \mathbb{N}$, $k > j$ implies $a_{k+p} = a_k$, then $\sum_{k=j+p+1}^{\infty} 10^{j+p-k} a_{k+p} = \sum_{k=j+1}^{\infty} 10^{j-k} a_k$.

3. Theorem. For each $k \in \mathbb{N}$, let $a_k \in \omega_0$ such that $a_k \leq 9$. Let $x = \sum_{k=1}^{\infty} 10^{-k} a_k$. If there exist $j \in \omega_0$, $p \in \mathbb{N}$ such that for all $k \in \mathbb{N}$, $k > j$ implies $a_{k+p} = a_k$, then $(10^{j+p} - 10^j)x = \sum_{k=1}^{j+p} 10^{j+p-k} a_k - \sum_{k=1}^{j} 10^{j-k} a_k$.

4. Prove Theorem 7.6.

**Notation.** We recall that a bar over a string of digits in the decimal expansion of a real number indicates that this string is repeated again and again without end. Thus the symbol $0.\overline{3}$ denotes the number $0.333\ldots$ The symbol $0.529\overline{871}$ denotes the number $0.5298718719871\ldots$ And so on.

5. Write the number $0.713\overline{29}$ in fractional form.

6. Write the number $0.\overline{16}$ in fractional form.

7. Write the number $0.\overline{142857}$ in fractional form.

8. Write the number $0.73\overline{9}$ in fractional form.

9. Write the number $0.74\overline{0}$ in fractional form. (Usually $0.74\overline{0}$ is written $0.74$. We have written $0.74\overline{0}$ to remind ourselves of the endless tail of zeroes.)

The following set of exercises will help us to prove our next major theorem.

**Exercises** (7.10)

1. Theorem. Let $n, m \in \mathbb{N}$ such that $n < m$. Let $r_0 = n$. For each $k \in \mathbb{N}$, let $q_k, r_k \in \omega_0$ such that $10r_{k-1} = q_k m + r_k$ and $r_k < m$. If there exists $j \in \mathbb{N}$ such that $r_j = 0$, then for all $k \in \mathbb{N}$, $q_{j+k} = 0$ and $r_{j+k} = 0$.

2. Theorem. Let $n, m \in \mathbb{N}$ such that $n < m$. Let $r_0 = n$. For each $k \in \mathbb{N}$, let $q_k, r_k \in \omega_0$ such that $10r_{k-1} = q_k m + r_k$ and $r_k < m$. If for all $1 \leq i \leq m-1$, $r_i \neq 0$, then there exist $j \in \omega_0$, $p \in \mathbb{N}$ such that $j + p < m$ and $r_j = r_{j+p}$. [Hint: Apply the pigeonhole principle to the sequence $\{r_k\}_{k=0}^{m-1}$.]

3. Theorem. Let $n$, $m \in \mathbb{N}$ such that $n < m$. Let $r_0 = n$. For each $k \in \mathbb{N}$, let $q_k$, $r_k \in \omega_0$ such that $10r_{k-1} = q_k m + r_k$ and $r_k < m$. If there exist $j \in \omega_0$, $p \in \mathbb{N}$ such that $r_j = r_{j+p}$, then for all $k > j$, $q_k = q_{k+p}$ and $r_k = r_{k+p}$.

4. Theorem. Let $n$, $m \in \mathbb{N}$ such that $n < m$. Let $r_0 = n$. For each $k \in \mathbb{N}$, let $q_k$, $r_k \in \omega_0$ such that $10r_{k-1} = q_k m + r_k$ and $r_k < m$. Then for each $k \in \mathbb{N}$, $0 \leq q_k \leq 9$.

5. Theorem. Let $n$, $m \in \mathbb{N}$ such that $n < m$. Let $r_0 = n$. For each $k \in \mathbb{N}$, let $q_k$, $r_k \in \omega_0$ such that $10r_{k-1} = q_k m + r_k$ and $r_k < m$. Then for each $t \in \mathbb{N}$, $\sum_{k=1}^{t} 10^{-k} q_k = \frac{n}{m} - \frac{10^{-t} r_t}{m}$.

6. Theorem. For each $k \in \mathbb{N}$, let $a_k \in \omega_0$ such that $a_k \leq 9$. If there exist $j \in \omega_0, p \in \mathbb{N}$ such that for all $k > j$, $a_k = a_{k+p}$, then $\sum_{k=j+p+1}^{\infty} 10^{p-k} a_k = \sum_{k=j+1}^{\infty} 10^{-k} a_k$. [Hint: Change the index of summation, using the substitution $i = k - p$.]

**Definition.** Let $x \in (0, 1)$, and let $\sum_{k=1}^{\infty} 10^{-k} a_k$ be a decimal expansion for $x$. If there exist $j \in \omega_0$, $p \in \mathbb{N}$ such that for all $k > j$, $a_k = a_{k+p}$, then the expansion $\sum_{k=1}^{\infty} 10^{-k} a_k$ is *periodic*.

**Definition.** Let $x \in (0, 1)$ such that $x$ has a periodic decimal expansion $\sum_{k=1}^{\infty} 10^{-k} a_k$. The *period* of the decimal expansion $\sum_{k=1}^{\infty} 10^{-k} a_k$ is the smallest number $p \in \mathbb{N}$ such that for some $j \in \omega_0$, for all $k \in \mathbb{N}$, if $k > j$ then $a_k = a_{k+p}$.

**Theorem (7.7)** For each $n$, $m \in \mathbb{N}$, if $n < m$ then the number $\frac{n}{m}$ has a periodic decimal expansion with period strictly smaller than $m$.

**Proof.** Let $n$, $m \in \mathbb{N}$, such that $n < m$. Let $r_0 = n$. For each $k \in \mathbb{N}$, let $q_k$, $r_k \in \omega_0$ such that $10r_{k-1} = q_k m + r_k$ and $r_k < m$.

Then (by Exercise 4) for each $k \in \mathbb{N}$, $0 \le q_k \le 9$. Hence there exists a number $x \in \mathbb{R}$ such that $0 \le x \le 1$ and for each $k \in \mathbb{N}$, $q_k$ is the $k$th digit after the decimal point in the decimal expansion of $x$. That is, there exists $x \in \mathbb{R}$ such that $0 \le x \le 1$ and $x = \sum_{k=1}^{\infty} 10^{-k} q_k$.

We will show that there exist $j \in \omega_0$, $p \in \mathbb{N}$ such that $j + p \le m - 1$ and $r_j = r_{j+p}$. Either there exists $j \in \mathbb{N}$ such that $j \le m - 1$ and $r_j = 0$, or not.

Case 1. Suppose that there exists $j \in \mathbb{N}$ such that $j \le m - 1$ and $r_j = 0$. Then (by Exercise 1) for all $k > j$, $r_k = 0$. Let $p = 1$. Then $p \le m - 1$ and $r_j = r_{j+p}$.

Case 2. Suppose that for all $1 \le j \le m-1$, $r_j \ne 0$. Then (by Exercise 2) there exist $j \in \omega_0, p \in \mathbb{N}$, such that $j + p \le m - 1$ and $r_j = r_{j+p}$. Since $j + p \le m - 1$ and $j \ge 0$, it follows that $p \le m - 1$.

Therefore, there exist $j \in \omega_0, p \in \mathbb{N}$ such that $p \le m - 1$ and $r_j = r_{j+p}$. It follows (by Exercise 3) that for all $k \in \mathbb{N}$, if $k > j$ then $q_k = q_{k+p}$. Therefore (by Theorem 7.6), the number $\sum_{k=1}^{\infty} 10^{-k} q_k$ is rational, and

$x = \frac{\sum_{k=1}^{j+p} 10^{j+p-k} q_k - \sum_{k=1}^{j} 10^{j-k} q_k}{10^j (10^p - 1)}$. Since $p < m$, the period of $\sum_{k=1}^{\infty} 10^{-k} q_k$ is strictly smaller than $m$.

Moreover (by Exercise 5), $\sum_{k=1}^{j+p} 10^{-k} q_k = \frac{n}{m} - 10^{-(j+p)} \frac{r_{j+p}}{m}$ and $\sum_{k=1}^{j} 10^{-k} q_k = \frac{n}{m} - 10^{-j} \frac{r_j}{m}$. Therefore, $\sum_{k=1}^{j+p} 10^{j+p-k} q_k - \sum_{k=1}^{j} 10^{j-k} q_k = 10^{j+p}(\frac{n}{m} - 10^{-j+p} \frac{r_{j+p}}{m}) - 10^j(\frac{n}{m} - 10^{-j} \frac{r_j}{m}) = (10^{j+p} - 10^j)(\frac{n}{m}) - \frac{r_{j+p}}{m} + \frac{r_j}{m}$. Since $r_{j+p} = r_j$, we have $\sum_{k=1}^{j+p} 10^{j+p-k} q_k - \sum_{k=1}^{j} 10^{j-k} q_k = (10^{j+p} - 10^j)(\frac{n}{m})$.

Thus, $x = \frac{10^j(10^p-1)(\frac{n}{m})}{10^j(10^p-1)} = \frac{n}{m}$.

It follows that $\frac{n}{m} = \sum_{k=1}^{\infty} 10^{-k} q_k$. Since there exist $j \in \omega_0$, $p \in \mathbb{N}$ such that $p \leq m-1$ and for all $k > j$, $q_k = q_k + p$, the decimal expansion $\sum_{k=1}^{\infty} 10^{-k} q_k$ is periodic with period strictly smaller than $m$.

Therefore, for all $n, m \in \mathbb{N}$, if $m < n$ then the number $\frac{n}{m}$ has a periodic decimal expansion, with period strictly smaller than $m$. Q.E.D.

**Exercises (7.11)**

1. Divide 1 by 17 on paper, using long division. What is the period of the decimal expansion you obtain?

2. Let $x \in (0,1)$ and suppose that $x$ has a periodic decimal expansion $\sum_{k=1}^{\infty} 10^{-k} a_k$. Let $t$ be the period of the expansion $\sum_{k=1}^{\infty} 10^{-k} a_k$. Then for each $p \in \mathbb{N}$ such that for some $j \in \mathbb{N}$, for all $k > j$, $a_k = a_{k+p}$, the period $t$ divides $p$.

3. Theorem. Let $m, n \in \mathbb{N}$, such that $n < m$. Let $r_0 = n$. For each $k \in \mathbb{N}$, let $q_k, r_k \in \omega_0$ such that $10r_{k-1} = q_k m + r_k$ and $r_k < m$. Then for each $i, k \in \mathbb{N}$, $r_{i+k} = 10^k r_i - \sum_{j=1}^{k} 10^{k-j} q_{i+j} m$. [Hint: Use induction on $k$.]

4. Theorem. For each $k \in \mathbb{N}$, $10^k - 1 = 9 \sum_{j=1}^{k} 10^{k-j}$.

5. Theorem. Let $m, n \in \mathbb{N}$ such that $n < m$. Let $r_0 = n$. For each $k \in \mathbb{N}$, let $q_k, r_k \in \omega_0$ such that $10r_{k-1} = q_k m + r_k$ and $r_k < m$. Then for each $i \in \mathbb{N}$ there exists $k \in \mathbb{N}$ such that $q_{i+k} \neq 9$. [Hints: Suppose that the conclusion is false. Then there exists $i \in \mathbb{N}$ such that for all $k \in \mathbb{N}$, $q_{i+k} = 9$. By the pigeonhole principle, there exist $s, k \in \mathbb{N}$ such that $s, k > i$, and $r_s = r_{s+k}$. Use the results of

Exercises 3 and 4 to show that $r_s = m$. Recall that for each $k \in \mathbb{N}$, $r_k < m$.]

6. Theorem. For each $x \in (0,1)$, there exists a decimal expansion $x = \sum_{k=1}^{\infty} 10^{-k} a_k$ such that for each $k \in \mathbb{N}$, for some $n \geq k$, $a_n \neq 9$. (We have already proved this in Chapter 5. But now we can prove it in another way.)

**Conclusion.** We believe that this is enough material for a one-semester first course in proving theorems. Therefore, with a bow to the reader and a hope that you have enjoyed the experience, we now conclude the main text of this book.

# Afterword A
# A Few Words on the History of Set Theory

*It is most unfortunate, but the point of this story has been reached where a justification of the expression "Murphy's mind" has to be attempted. Happily we need not concern ourselves with this apparatus as it really was — that would be an extravagance and an impertinence — but solely with what it felt and pictured itself to be. Murphy's mind is after all the gravamen of these informations. A short section to itself at this stage will relieve us from the necessity of apologizing for it further.*

*Samuel Beckett,* Murphy

This book introduces modern set theory and modern mathematical notation. These form the basic language and conventions of modern mathematics. We regard them as given, as we regard modern English as given. They are the medium in which we communicate.

But mathematical discourse was not always as it is now (Nor, for that matter, was English.). Anything that seems difficult in modern notation used to be much worse. Zero, Arabic numerals, fractions, decimals, the equal sign, negative numbers, the "mod $n$" notation, the summation sign, the modern definitions of "relation" and "function": all of these modern conveniences make it easier to discuss mathematics. Similarly, set theory and modern conventions of proof make it much easier to prove theorems and to read and evaluate proofs. We need to be able to tell whether or not a proof is correct. If we can't tell, then the proof is not yet clearly written, not yet clearly understood.

Mathematics is a natural activity of the human imagination. People invent mathematics first and explain it afterwards. Sometimes what we discover doesn't seem to make sense to us. The Pythagoreans could demonstrate to themselves with terrifying ease that the square root of two was irrational, but they did not like the idea of irrational numbers. An irrational number essentially involves the notion of infinity.

In this book we define the real numbers only as formal symbols. A positive real number $x$ can be written as a decimal, which we can write as $\sum_{k=0}^{\infty} a_k 10^{-k}$, where $a_0$ is a positive integer and for all $k \in \mathbb{N}$, $a_k$ is a one-digit nonnegative integer. We do not discuss the real meaning of the notation $\sum_{k=0}^{\infty} a_k 10^{-k}$, which is $\lim_{n\to\infty} \sum_{k=0}^{n} a_k 10^{-k}$. This would take us too far afield, into a discussion of limits and convergence. (See Afterword B.) This is traditionally the province of the subject known as real analysis. Hence we have omitted it from our introduction to set theory and proof. We regard the real numbers naively, the way we have been taught in school to regard them, as infinite strings of digits. But what are the real numbers, really?

This is the sort of question that has always bothered both mathematicians and non-mathematicians. Is there really such a thing as a negative number? Are there really infinitely many positive integers? Do irrational numbers really exist? What about imaginary numbers? Can negative numbers have square roots?

In the last half of the 19th century, mathematicians were trying to figure out how to define limits. Calculus, the language of Newton's physics, had existed for over a hundred years, and calculus essentially involves limits. People explained limits in terms of what they called infinitesimals — numbers smaller than any positive rational number but larger than zero. In mainstream modern mathematics, infinitesimals do not exist. (Since an infinitesimal number expresses a magnitude between 0 and 1, an infinitesimal should be a positive real number. For there are no "gaps" in the real numbers. Real numbers are the only numbers which express finite sizes. Mainstream modern analysis accepts the Axiom of Archimedes, which states that for every $\varepsilon \in \mathbb{R}$, if $\varepsilon > 0$ then for each $n \in \mathbb{N}$, there exists $k \in \mathbb{N}$ such that $k\varepsilon > n$. Thus, there are no infinitesimals in mainstream modern analysis. There *are* modern mathematicians who work with infinitesimals, to see how much can be done with them. Analysis with infinitesimals is called non-standard analysis.) The modern theory of limits is the subject of real analysis, the mathematics which underlies calculus. The student who has experienced and done well with the material in this book is in a good position to study real analysis, as well as number theory, modern algebra, and topology.

Modern analysis and topology were in the process of being invented in the late 19th century. Mathematicians sought to explain infinitesimals and define the real numbers to their satisfaction.

Part of the reason for this surge of interest in certainty about the real numbers was the discovery (by Lobachevsky, among others), during the first half of the 19th century, that Euclid's parallel postulate (that given a point $p \in \mathbb{R}^3$ and a line $L_1 \subseteq \mathbb{R}$ such that $p_1 \notin L_1$, there exists exactly one line $L_2 \subseteq \mathbb{R}^3$ such that (1) $p \in L_2$, (2) $L_2$ is coplanar with $p$ and $L_1$, and (3) $L_1 \cap L_2 = \varnothing$) is not true in all possible geometries. In hyperbolic geometry, for example, there are infinitely many coplanar lines parallel to a given line through a given point not on the line. Although the world around us looks more Euclidean than hyperbolic, hyperbolic geometry is as "real" as Euclidean geometry. (Complex numbers are naturally related to hyperbolic geometry.) In spherical geometry, parallel lines do not exist.

Mathematicians found it astonishing, and also rather alarming, that Euclid's parallel postulate was not a self-evident fact and part of the laws of thought. If other geometries can exist, then what else might not exist? And if our "natural" geometric intuition can't be trusted, then what *can* be trusted? Thus, mathematicians at this time felt creatively inspired to invent new mathematics, and at the same time they were fired with the desire to test the certainty of established mathematics. Calculus, irrational numbers, the theory of limits, and infinity in general were all crying out for a closer look.

It was in this atmosphere that Georg Cantor invented his *Mengenlehre*, or set theory (first published in 1895). Cantor gave an explicit definition of a set as a collection of objects. If you have worked through this book, you are familiar by now with the role of definitions in mathematical proof. A mathematician is entitled to take a definition and work with it, word by word, to wring conclusions from it. Not every explanatory phrase will stand up to such treatment.

One problem with Cantor's definition of "set" is that it allows one to imagine and discuss the set of all sets, the set whose elements are all the sets that exist.

Bertrand Russell publicly discredited Cantor's set theory by publishing what came to be known as the Russell Paradox. Really, the Russell Paradox is the proof of a theorem, not a paradox. Here is the proof.

**Theorem.** There is no such thing as the set of all sets.

**Proof.** By way of contradiction, suppose that there exists a set $G$ such that for each set $A$, $A \in G$. Then $G \in G$.

Let $H$ be the set $\{S \in G \mid S \notin S\}$. Either $H \in H$ or $H \notin H$.

Case 1. Suppose that $H \in H$. Then, by definition of $H$, $H \notin H$. Thus $H \in H$ and $H \notin H$. $\rightarrow\leftarrow$ .

Case 2. Suppose that $H \notin H$. Then, by definition of $H$, $H \in H$. Thus $H \notin H$ and $H \in H$. $\rightarrow\leftarrow$

The hypothesis that the set $G$ exists leads to a contradiction. Therefore, this hypothesis is false. The set $G$ does not exist. There is no set of all sets. Q.E.D.

This proof looks fishy, and, indeed, it is not quite correct in modern set theory. If we are going to consider the possibility of a "set of all sets," we should be careful to respect the modern rule that, for each proof, there must exist a universe of discourse $X$ such that for each set $A$ mentioned in the proof, $A \subseteq X$. With this in mind, let's look at a modern version of the Russell Paradox.

**Theorem.** There is no such thing as the set of all sets.

**Proof.** Let $X$ be a universe of discourse. By way of contradiction, suppose that there exists a set $G$ of all sets. Since $X$ is our universe of discourse, $G$ is the set of all subsets of $X$. That is, $G = \mathcal{P}(X)$. Since $G$ is a set and $X$ is our universe of discourse, $G \subseteq X$. Thus, $\mathcal{P}(X) \subseteq X$.

Let $A \in \mathcal{P}(X)$, and let $f\colon X \to \mathcal{P}(X)$ be defined by

$$f(y) = \begin{cases} y, \text{ if } y \in \mathcal{P}(X); \\ A \text{ otherwise.} \end{cases}$$

Let $S \in \mathcal{P}(X)$. Since $\mathcal{P}(X) \subseteq X$ and $S \in \mathcal{P}(X)$, $S \in X$. Since $S \in X$ and $S \in \mathcal{P}(X)$, $f(S) = S$. Thus, for each $S \in \mathcal{P}(X)$, there exists $x \in X$ (namely $x = S$) such that $f(x) = S$. That is, $f$ is onto.

Let $H = \{y \in X \mid y \notin f(y)\}$. Since $H \in \mathcal{P}(X)$, and since $f$ is onto, there exists $z \in X$ such that $f(z) = H$. Either $z \in H$ or $z \notin H$.

Case 1. Suppose that $z \in H$. Then, by definition of $H$, $z \notin f(z)$. Since $z \notin f(z)$ and $f(z) = H$, $z \notin H$. Thus $z \in H$ and $z \notin H$. $\rightarrow\leftarrow$

Case 2. Suppose that $z \notin H$. Then, by definition of $H$, $z \in f(z)$. Since $z \in f(z)$ and $f(z) = H$, $z \in H$. Thus $z \notin H$ and $z \in H$. $\rightarrow\leftarrow$

Our hypothesis has led to a contradiction and is therefore false. Therefore, for each universal set $X$, $\mathcal{P}(X) \not\subseteq X$. That is, the set of all sets does not exist. Q.E.D.

Thus, what the Russell Paradox actually proves is that for each set $X$, $\mathcal{P}(X) \not\subseteq X$.

For each set $X$, $\mathcal{P}(X)$ is a set. But since $\mathcal{P}(X) \not\subseteq X$, $X$ cannot be a universe of discourse for a proof involving $\mathcal{P}(X)$. There is no universe of discourse that actually contains all sets.

This is how we talk now, with hindsight. At the end of the 19th century, though, Russell reasoned more like this. A set is any collection of objects. Therefore, the set of all sets is a set. Let $X$ be the set of all sets. Let $H = \{S \in X \mid S \notin S\}$. Then $H$ is a set, so $H \in X$. This leads to a contradiction. Therefore, there is something wrong with the definition of "set" as "any collection of objects."

Russell tried to repair Cantor's set theory with a "theory of types" to ensure that no set could belong to itself. (With hindsight, we can see that the contradiction in the Russell Paradox comes not from the notion of a set's belonging to itself, but from the idea of a set which has its own power set as a subset.) Together with Alfred North Whitehead, Russell wrote *Principia Mathematica*, a gigantic failed attempt to reduce all of mathematics to set theory and logic.

Cantor's set theory included infinite numbers of two kinds: ordinal numbers (which can be used to index infinite sets) and cardinal numbers (used for the cardinalities of infinite sets). This theory, with slight modifications, is still used today. If you take a course or work your way through a book on advanced set theory, you will learn about transfinite ordinal and cardinal numbers. In point-set topology, each ordinal number is defined as the set of ordinal numbers strictly smaller than itself. Thus $0 = \{\ \}$, $1 = \{0\}$, $2 = \{0, 1\}$, and so on. You may recall that $\omega_0 = \{0, 1, 2, 3, \ldots\}$. Thus, $\omega_0$ (omega nought) is the smallest infinite ordinal number. As a cardinal number, $\omega_0 = |\omega_0| = |\mathbb{N}|$. Modern set theory, following Cantor, defines addition, multiplication, exponentiation, and so on, of transfinite numbers. It turns out that $2^{\omega_0} = |\mathcal{P}(\mathbb{N})| = |\mathbb{R}|$, and $|\mathcal{P}(\mathcal{P}(\mathbb{N}))| = 2^{2^{\omega_0}}$, and $|\mathcal{P}(\mathcal{P}(\mathcal{P}(\mathbb{N})))| = 2^{2^{2^{\omega_0}}}$. Thus, there exist infinitely many infinities of various sizes.

Infinities of different sizes struck many people as a crazy idea. Moreover, Cantor himself was a mystic, and considered these infinities to be related to the Absolute, or God. Cantor's religious beliefs in no way affect the validity of his mathematics, but they did influence the way his contemporaries viewed his mathematics. Some people thought that Cantor was twisting mathematics in order to satisfy his mysticism. Also, unsurprisingly, Cantor made a few mistakes. New and original mathematics is seldom completely perfect in its first form. When new mathematics is good, as Cantor's was, it gets worked over by many people, and the flaws in it are repaired. The first step in this process, the finding and publishing of the holes in the argument, can be painful for the pioneer whose work is being picked apart. It was painful for Cantor.

Various "paradoxes of set theory" were invented on the model of the Russell Paradox. (Littlewood refers to the paradoxes as "splendid jokes.") One of the best (Richard's Paradox) concerns "the least number unnameable in English in under twenty syllables." The phrase in quotation marks has fewer than twenty syllables, and defines the number in question. (Moreover, once we say: "Let $n$ be the least integer unnameable in English in under 20 syllables," there is the difficulty that now the number $n$ can be named in one syllable.) Can a person who claims to be lying really be lying? Can a person who claims to be lying be telling the truth? Is the statement "I am lying" true, false, or neither?

Meanwhile, Cantor's uncountable infinities were proving extremely useful to the mathematicians (such as Poincaré and Weierstrass) inventing topology and analysis. While some mathematicians (such as Kronecker) found set theory crazy and nonsensical sounding, others found that it was just what they needed. With structures such as topology and measure theory resting on them, mathematicians became very loath to part with the new infinities.

One of the mind-boggling features of Cantor's infinities of different sizes is that their existence is so easy to prove. Theorem 5.10 in this book, stating that there is no onto function from a set to its own power set, suffices to show that there are many different sizes of infinity. The proof of this theorem is short, and resembles the proof known as the Russell Paradox. Cantor's famous diagonal argument, Theorem 5.17 in this book, proves that the open interval $(0, 1)$ is uncountable. Again, the proof is simple. The diagonal argument is often successfully presented in philosophy classes to students who know very little mathematics. How can such an important, astonishing theorem be so easy to prove? The feeling of some of Cantor's contemporaries was that this simple proof with the bizarre conclusion must be a trick.

Another mysterious feature of the new set theory was the continuum hypothesis. Let $\omega_1$ denote the least uncountable ordinal number. As an ordinal number, $\omega_1$ is the set of all ordinals smaller than $\omega_1$. As a cardinal number, $\omega_1 = |\omega_1|$. Recall that $|\mathbb{N}| = \omega_0$ and $|\mathbb{R}| = 2^{\omega_0}$. Is $\omega_1$ equal in cardinality to $2^{\omega_0}$? Or is $|\omega_1| < |2^{\omega_0}|$? Does there exist an infinite set $S \subseteq \mathbb{R}$ such that $|S| \neq |\mathbb{N}|$ and $|S| \neq |\mathbb{R}|$? The continuum hypothesis was the conjecture that $\omega_1 = 2^{\omega_0}$. Cantor first made this conjecture around 1880. Neither he nor anyone else succeeded in either proving or disproving it until 1963. In that year, Paul J. Cohen proved that the continuum hypothesis is independent of the other axioms of set theory, just as Euclid's parallel postulate is independent of the other axioms of Euclidean geometry. Thus, the continuum hypothesis can be neither proved nor disproved. A consistent system of axiomatic set theory exists in which the continuum hypothesis is true, and another consistent system exists in which the continuum hypothesis is false. Since this was not proved until 1963, the continuum hypothesis was just one more muddle and source of confusion at the end of the 19th century.

Thus, Cantor's set theory excited a great deal of controversy. Some people wanted to prove set theory correct, and some wanted to prove it wrong. Mathematicians (von Neumann, Bernays, Gödel, Zermelo, Peano, and others) constructed systems of axioms to clarify and justify set theory. There are several such systems which seem to be correct. However, whenever mathematicians try to prove that a particular axiom system is *exactly* the one we need, or that all of mathematics can be reduced to logic, things go awry. In 1930, Kurt Gödel proved that every formal system rich enough to contain recursion (or induction, or the set of natural numbers) contains truths that cannot be proved in the system. People found this dismaying, but, with hindsight, it seems no more surprising than the impossibility of proving the existence or nonexistence of God from formal logic alone. We can't prove everything, and we can't prove that our system does not allow us to prove false statements. We can't prove within mathematics that mathematics is somehow "absolutely correct." We can't prove once and for all that we really know everything we think we know. Perhaps this should not surprise us, but in general people do, and historically did, find it shocking at first. (Gödel himself was not shocked by his conclusion. He was dismayed by attempts to prove that mathematics was nothing but logic, and could be done by machine. Gödel felt that his conclusion showed that human intuition was necessary in mathematics, and that the human mind was not a computer.)

Controversy ensued concerning the nature of mathematics. Some people, known as constructivists, declared that infinity does not exist, that there is no such thing as the set $\mathbb{N}$. (You can see their point. You will never be able to list all the elements of $\mathbb{N}$, so how can you claim to know that they exist?) Most constructivists reject the Law of Excluded Middle ($\neg\neg p \to p$), because they regard the statement "The proposition $p$ is true," as equivalent to, "The proposition $p$ has been proved directly." They reject proofs by contradiction. Formalists regard mathematical formulas and proofs as strings of symbols subject to rules. It is fruitless, they say, to worry about the meanings of the symbols. (You can see the formalists' point, too. They get tired of vague questions about the nature of infinity and the uncanniness of natural numbers. They want to get on with the mathematics.) Platonists regard numbers as real, existing entities, as children do. (Children often think of numbers as people.) Most mathematicians talk like Platonists most of the time.

There are several other possible and actual standpoints for philosophers of mathematics. Abraham Frankel and Yehoshua Bar-Hillel, in their entertaining history *Foundations of Set Theory*, remark that mathematicians who "dabble in philosophy" seldom stick to one philosophical position. For those of us who are not specialists in logic or mathematical foundations, the various philosophical schools seem somewhat like moods that every mathematician falls into, with the Platonic mood predominating. Many mathematicians had the numbers as imaginary friends in childhood. They are among our oldest friends, so of course we are naively inclined to regard them as existing.

Nobody really knows what mathematics is. It is something that people do. Mathematics is widely regarded as useful for understanding "the real world," but this usefulness is among the things that can't be proved.

Most universities offer a course called Philosophy of Mathematics, usually in the Department of Philosophy. This course presents some of the history of early set theory. It's a very entertaining course.

Modern analysis and topology make enthusiastic use of Cantor's uncountable infinities. Analysis, topology, number theory and modern algebra are fascinating subjects that will not disappoint you. If you have worked through this book, you are ready to study them.

Several axiomatic systems for set theory have been devised. They are all satisfactory for most mathematical purposes, but no system will do everything desired by those who work on the cutting edge of set theory and logic.

There will always be some mist shrouding the foundations of mathematics. We will never be able to define the word "set." (We could, of course, define the word "set," but only at the expense of introducing some other undefined word.) And there will always be mathematicians working at the edge of intelligibility, on logic, computability, very rapidly growing functions, and so on. These lie outside the scope of this book, which presents only the set theory that "everybody knows." To learn more, read other books and take other classes. The wide, weird world of mathematics awaits you.

# Afterword B
# A Little Bit About Limits

We do not discuss limits in the main text of this book. But we do use the notation $\sum_{k=1}^{\infty} 10^{-k}a_k$ for the number which can be written as a decimal point followed by the infinite string of digits $a_1, a_2, a_3 \ldots a_k \ldots$. And we have mentioned that $\sum_{k=1}^{\infty} 10^{-k}a = \lim_{n\to\infty} \sum_{k=1}^{n} 10^{-k}a_k$. We feel that a few more explanatory remarks are in order, for the sake of the conscientious reader. But the discussion here will be very incomplete. The subject of limits and irrational numbers is a large one. To really understand limits, you will need to spend a full semester studying real analysis or advanced calculus. We can't do the subject justice in an afterword.

**Definition.** Let $\{x_k\}_{k=1}^{\infty}$ be a sequence of real numbers. If there exists $L \in \mathbb{R}$ such that for each real number $\varepsilon > 0$, there exists $n \in \mathbb{N}$ such that for all $k \in \mathbb{N}$, if $k \geq n$ then $|x_k - L| < \varepsilon$, then the sequence $\{x_k\}_{k=1}^{\infty}$ *converges* to the number $L$, and the number $L$ is a *limit* of the sequence $\{x_k\}_{k=1}^{\infty}$. The notation $\lim_{k\to\infty} x_k = L$ means that the sequence $\{x_k\}_{k=1}^{\infty}$ converges to $L$.

**Axiom** (Axiom of Archimedes). For each $\varepsilon \in \mathbb{R}$ if $\varepsilon > 0$ then for each $n \in \mathbb{N}$, there exists $m \in \mathbb{N}$ such that $m\varepsilon > n$.

**Theorem 1.** For each $n \in \mathbb{N}$, $2^n > n$.

**Proof.** Exercise.

**Theorem 2.** For each $\varepsilon \in \mathbb{R}$, if $\varepsilon > 0$ then there exists $n \in \mathbb{N}$ such that $\frac{1}{2^n} < \varepsilon$.

**Proof.** Exercise.

**Theorem 3.** For each $k \in \mathbb{N}$ let $x_k = \frac{2^k-1}{2^k}$. (Thus $x_1 = \frac{1}{2}$, $x_x = \frac{3}{4}$, $x_3 = \frac{7}{8}$, and so on.) Then $\lim_{k \to \infty} x_k = 1$.

**Proof.** Let $\varepsilon \in \mathbb{R}$ such that $\varepsilon > 0$. Then, by Theorem 2, there exists $n \in \mathbb{N}$ such that $\frac{1}{2^n} < \varepsilon$. Let $k \geq n$. Since $x_k = \frac{2^k-1}{2^k}$, $|x_k - 1| = \frac{1}{2^k}$. Since $k \geq n$, $\frac{1}{2^k} \leq \frac{1}{2^n} < \varepsilon$. Thus, for each $k \in \mathbb{N}$, if $k \geq n$ then $|x_k - 1| < \varepsilon$.

Since there exists $n \in \mathbb{N}$ such that, for each $k \in \mathbb{N}$, if $k \geq n$ then $|x_k - 1| < \varepsilon$, it follows that $\lim_{k \to \infty} x_k = 1$. Q.E.D.

**Definition.** Let $\{x_k\}_{k=1}^{\infty}$ be a sequence of real numbers. If $\{x_k\}_{k=1}^{\infty}$ does not converge, then $\{x_k\}_{k=1}^{\infty}$ *diverges.*

**Theorem 4.** For each $k \in \mathbb{N}$, let $x_k = (-1)^k$. (Then $x_1 = -1$, $x_2 = 1$, $x_3 = -1$, and so on.) The sequence $\{x_k\}_{k=1}^{\infty}$ diverges.

**Proof.** By way of contradiction, suppose that the sequence $\{x_k\}_{k=1}^{\infty}$ converges. Then there exists $L \in \mathbb{R}$ such that for each real number $\varepsilon > 0$, there exists $n \in \mathbb{N}$ such that for all $k \in \mathbb{N}$ if $k \geq n$ then $|x_k - L| < \varepsilon$.

Let $\varepsilon = \frac{1}{3}$. Then there exists $n \in \mathbb{N}$ such that for each $k \in \mathbb{N}$, if $k \geq n$ then $|x_k - L| < \frac{1}{3}$.

Let $k$ be an odd positive integer such that $k \geq n$. Then $x_k = -1$ and $x_{k+1} = 1$. Since $k \geq n$, $|x_k - L| < \frac{1}{3}$ and $|x_{k+1} - L| < \frac{1}{3}$. Since $|x_k - L| < \frac{1}{3}$, it follows that $|-1 - L| < \frac{1}{3}$. Thus $\frac{-2}{3} > L > \frac{-4}{3}$. Hence $L < 0$. Since $|x_{k+1} - L| < \frac{1}{3}$, it follows that $|1 - L| < \frac{1}{3}$. Thus $\frac{4}{3} > L > \frac{2}{3}$. Hence $L > 0$. Thus $L < 0$ and $L > 0$. $\rightarrow\leftarrow$

Our hypothesis has led to a contradiction and is therefore false. Therefore, the sequence $\{x_k\}_{k=1}^{\infty}$ diverges. Q.E.D.

**Definition.** Let $\{c_k\}_{k=1}^{\infty}$ be a sequence of real numbers. For each $n \in \mathbb{N}$ let $x_n = \sum_{k=1}^{n} c_k$. Then the *series* $\sum_{k=1}^{\infty} c_k$ is the sequence $\{x_n\}_{n=1}^{\infty}$.

**Definition.** Let $\sum_{k=1}^{\infty} c_k$ be a series of real numbers. For each $n \in \mathbb{N}$ let $x_n = \sum_{k=1}^{n} c_k$. The series $\sum_{k=1}^{\infty} c_k$ *converges* to a limit $L \in \mathbb{R}$ if $L = \lim_{x \to n} x_n$. If the series $\sum_{k=1}^{\infty} c_k$ does not converge to a limit, then the series $\sum_{k=1}^{\infty} c_k$ *diverges.* To express the statement that the series $\sum_{k=1}^{\infty} c_k$ converges to a limit $L \in \mathbb{R}$, we write $\sum_{k=1}^{\infty} c_k = L$.

**Theorem 5.** For each $n \in \mathbb{N}$, $\sum_{k=1}^{n} \frac{1}{2^k} = \frac{1-2^n}{2^n}$.

**Proof.** Exercise.

**Theorem 6.** $\sum_{k=1}^{\infty} \frac{1}{2^k} = 1$.

**Proof.** Exercise.

**Theorem 7.** Let $c_1 = -1$, and for each $k \in \mathbb{N}$, if $k > 1$ then let $c_k = (-1)^k(2)$. Then for each $n \in \mathbb{N}$, $\sum_{k=1}^{n} c_k = (-1)^n$.

**Proof.** Exercise.

**Theorem 8.** Let $c_1 = -1$, and for each $k \in \mathbb{N}$, if $k > 1$ then let $c_k = (-1)^k(2)$. Then $\sum_{k=1}^{\infty} c_k$ diverges.

**Proof.** Exercise.

**Theorem 9.** Let $\{x_k\}_{k=1}^{\infty}$ be a sequence of real numbers. If there exist $L, M \in \mathbb{R}$ such that $\lim_{k \to \infty} x_k = L$ and $\lim_{k \to \infty} x_k = M$, then $L = M$.

**Proof.** Exercise.

Let's return to the real numbers. We recall that the set of rational numbers is $\{\frac{a}{b} \mid a \in \mathbb{Z} \text{ and } b \in \mathbb{N}\}$. We do not require any infinite process to understand the rational numbers.

The rational numbers do not form a continuum; there are "gaps" in the set $\mathbb{Q}$. For example, consider the sets $A = \{x \in \mathbb{Q} \mid x^3 \leq 7\}$ and $B = \{x \in \mathbb{Q} \mid x^3 \geq 7\}$.

For each $x \in A$, $\sqrt[3]{7} > x$, and for each $x \in B$, $\sqrt[3]{7} < x$. Moreover, for each $y \in \mathbb{R}$, if for each $z \in A$, $y > z$, and if for each $v \in B$, $y < v$, then $y = \sqrt[3]{7}$. This gives a way to specify the irrational number $\sqrt[3]{7}$. Thus, one way to define irrational numbers is by considering "gaps" or "cuts" in the rationals.

Another way to specify an irrational number is as the limit of a convergent sequence of rationals. But, to do this without circularity, we must have some notion of convergence for sequences that does not mention the limit $L$. This gets too complicated for us to describe further here, but it is done near the beginning of a real analysis course.

In this book, we think of any real number $x$ between 0 and 1 in terms of an infinite sequence $\{a_k\}_{k=1}^{\infty}$ preceded by a decimal point. This is the same thing as thinking of $x$ as the infinite sum $\sum\limits_{k=1}^{\infty} 10^{-k} a_k$, or as $\frac{a_1}{10} + \frac{a_2}{100} + \frac{a_3}{1000} + \ldots + \frac{a_k}{10^k} + \ldots$, or as the limit of the sequence $\{\sum\limits_{k=1}^{n} 10^{-k} a_k\}_{n=1}^{\infty}$ of rational numbers. But are we really sure that the number $x$ exists, that it really makes sense to speak of such a number? And if it doesn't make sense, then why not?

In a course in real analysis (or advanced calculus), we work extensively with the definitions of limit and convergence. We define open sets, closed sets, limit points of sets, and so on. We get a much more detailed understanding of the real numbers. We will not discuss it further here, not because it is too difficult, but because there is too much to explain. We refer you to a course or a book on real analysis. (See Afterword D.) It's a wonderful subject, and will not disappoint you.

# Afterword C
# Why No Answers In The Back of the Book?

Unlike most modern undergraduate mathematics textbooks, this book does not include a section with answers to all or some of the problems. There are several reasons for this.

In the first place, most of the solutions to the problems in this book are proofs. There is not just one correct answer to most of these problems. But if answers are supplied, many students will regard the book's answers as automatically superior to their own. Answers in the back of the book can result in a depressing uniformity in students' papers.

Secondly, the point of this course is not that the student learn a certain body of information about sets, but that the student learn the process of constructing and writing proofs. Looking up answers is not part of the process of proving theorems. But looking up answers is addictive. We have heard even quite talented students say that doing problems is useless unless someone gives you the answer.

This is contrary to the whole spirit of mathematics. Solving a problem in mathematics does not end with getting the answer. The real mathematics *starts* once you have an answer. The real mathematics consists of proving that your answer is correct (and, in case your answer is in in discovering and confirming this fact). Answers in the back of the book weaken students.

Often students believe that answers in the back of the book will help them to get better grades. In fact, the opposite is true. Students who rely heavily on precooked answers tend to do badly on exams, where no such answers are given. The answers in the back of the book may possibly boost the students' homework grades, but they tend to lower exam grades.

Moreover, it's not a good idea for students to boost their homework grades artificially. The professor looks at your homework to figure out what you know and what you don't know, with a view to helping you learn. And the professor designs the exams based on the level of knowledge indicated by the students' homework. If your homework shows that you know much more than you actually do know, you are likely to wind up taking exams that are much too hard for you. This will not improve your grade. Nor will it teach you mathematics.

If you are worried because your homework grade is low, try asking the professor for extra problems. Doing extra homework problems tends to improve students' exam grades, even in cases where the homework grade does not improve.

The mathematics in this book is verbal mathematics, the mathematics of proof. More words than numbers are used. The point of this course is for the student to begin to use and understand mathematical language and to be able to write proofs. For most students, this is a new sort of learning. You may well feel at first as if you are not very good at it. (Oddly enough, very talented students are much more likely to feel this way than poor students. Good students tend to be underconfident, and poor students overconfident.) One good thing about mathematics is that it is inherently learnable. If you keep at it you will learn. Some learn faster than others, but everyone learns. It can be exasperating to be one of the slower learners, but please do remember that many factors affect your learning speed: your background, the amount of time you can devote to studying, and so on. Learning more slowly than someone else does not mean that you will never learn the subject well. It merely means that the process will take longer for you than for some other person. This is nothing to be ashamed of. It is the student who increases her/his knowledge most, rather than the student who reaches the highest knowledge level, who really gets the most out of the class. If you're slow, you will probably not get the highest grade in this particular class. But once you finally learn the process of writing proofs, you will get high grades in other classes.

The students who do poorly in this class are usually those who are too complacent. They attend class less frequently than other students, and regard a steady 50% on the homework as pretty good. They turn in less homework than the others, but are surprised when they do poorly on exams. When we have review sessions, the poor students (who really need the review) often fail to attend, while the good students are glad of the review (although they seem to need it much less).

You don't need answers in the back of the book. You can be strong and write your own proofs. Learn to reason effectively, to communicate your reasoning, and to rely on your own efforts to test and correct your conclusions. If you can learn this, you'll find that you have become "good at math," and have no need of answers in the back of the book.

# Afterword D: What next? Concise Synopses of Selected College Mathematics Courses

*Said Conrad Conelius o'Donald o'Dell,*
*My very young friend who is learning to spell:*
*"The A is for Ape. And the B is for Bear.*
*The C is for Camel. The H is for Hare.*
*The M is for Mouse. And the R is for Rat.*
*I know* all *the twenty-six letters like that...*

*"...through to Z is for Zebra. I know them all well."*
*Said Conrad Cornelius o'Donald o'Dell.*
*"So now I know everything* any*one knows*
*From beginning to end. From the start to the close.*
*Because Z is as far as the alphabet goes."*

*Then he almost fell flat on his face on the floor*
*When I picked up the chalk and drew one letter more!*
*A letter he never had dreamed of before!*
*And I said, "*You *can stop, if you want, with the Z*
*Because most people stop with the Z*
But not me!

*"In the places I go there are things that I see*
*That I* never *could spell if I stopped with the Z.*
*I'm telling you this 'cause you're one of my friends.*
My *alphabet starts where* your *alphabet ends!"*

*Dr. Suess,* On Beyond Zebra

Now that you have begun to learn the language of modern mathematics, you may well want to take a few more mathematics courses. What follows is a list of brief descriptions of a few typical undergraduate mathematics courses. Most of the courses we describe are verbal courses which emphasize the writing of proofs, but we also include Calculus and Probability. There are many good classes that we are not describing. See what your local universities have to offer.

1. **Real Analysis.** This class is also called Advanced Calculus. (Advanced Calculus means different things at different universities.) In this course you will use the definition of "limit" extensively. You will also define the real numbers, and prove theorems about convergence of sequences to limits, uniform convergence, continuous functions, compact sets, and so on. Toward the end of the course you will define the Riemann integral and the derivative of a function. You will also make the acquaintance of the standard middle-thirds Cantor set on the real line. After you take this course, you will be able to write and read proofs on a level with those that appear in calculus books. It is not necessary to take calculus before real analysis, although most people do so. This course will make calculus make sense to you.

2. **Calculus.** The first year of calculus is not usually taught as a verbal, proof-oriented course. Proofs appear in the text and some theorems are proved by the instructor. The student learns to compute limits, differentiate functions, and evaluate integrals. Interesting and ingenious methods of calculation are introduced. Some people love first-year calculus, some tolerate it, and some hate it. If you are

a very "verbal" person, you may wish to study real analysis before calculus. Once you have learned real analysis, in fact, you can, if necessary, teach yourself calculus.

All mathematicians know calculus, so it's important to learn it. Calculus can tell you many things about functions. Also, you need to know calculus well to really appreciate probability. Nearly all of the less verbal, more computational mathematics courses (which are fascinating in their own right, and useful in physics) require calculus as a starting point.

3. **Probability.** Probability is not usually taught as a proof course, but it uses verbal descriptions and set theory. Probability is usually taught in terms of story problems. Understanding the problems involves reasoning and set theory. Solving them requires combinatorics and calculus. The student who has done well with the material in this book, and who also knows calculus, will be able to understand and appreciate the proofs in the probability course. Since probability describes gambling, it is inherently fascinating. The results are often surprising. Probability is a lot of fun.

4. **Linear Algebra.** Courses in linear algebra go under several different names. Matrices, linear transformations, and vector spaces are the chief objects of study. There are two sorts of linear algebra course. One is proof-oriented, and involves proving theorems about vector spaces. This is usually considered a good proof class for beginners, so it can be a good class to take after finishing this book. The other sort of linear algebra class is more computational and may cover more applications. You don't need calculus to study linear algebra, although calculus is often listed as a course prerequisite.

5. **Elementary Number Theory.** Number theory is about the natural numbers: divisibility, properties of prime numbers, distribution of primes, solutions to equations modulo a natural number $n$, and so on. The natural numbers are surprising. Number theory is usually taught as a proof course, and is a good course to take after finishing this book. Calculus is not needed for elementary number theory.

6. <u>**Modern Algebra.**</u> Modern algebra is somewhat difficult to describe, because it is really quite different from high school algebra, calculus, or anything you have seen before. We study groups, rings, and fields: sets equipped with operations. It's a new way of thinking, full of fascinating puzzles. The contents of this book are a good backround for modern algebra. You do not need calculus to take this course.

   Because modern algebra is the general study of sets with operations, it is more abstract than linear algebra and number theory. It is a beautiful subject, and if you understand the material in the first five chapters of this book, you will enjoy modern algebra immensely.

7. <u>**Advanced Set Theory.**</u> This course covers the more infinite portions of set theory, including transfinite ordinals and cardinals. Basically, we count to infinity and just keep counting. To be more precise, we can't really count up to the ordinal $\omega_0$ (omega nought), the first infinite ordinal. Hence $\omega_0$ is called a *limit ordinal.* But $\omega_0+1$ is a *successor ordinal.* The distinction between countable and uncountable remains important. In this course you will meet the axiom of choice, the maximal principal, Zorn's lemma, transfinite induction, and so on. All of this set theory is used in higher-level analysis, algebra, topology, and other branches of mathematics.

8. <u>**Topology.**</u> Usually the first course in topology is an introduction to general, or point-set, topology. A topological space $X$ is a set $X$, together with a collection of subsets of $X$ which satisfies certain axioms. Point-set topology concerns open sets, limit points, and continuous functions. It can be described as an abstract study of spaces, and has great appeal for people who think in pictures rather than in words. Topology is usually taught as a proof class, with the students proving theorems at the blackboard. Drawing pictures with colored chalk is often encouraged. No calculus is needed.

9. **Analysis or Calculus with Complex Variables.** This may or may not be taught as a proof class. If you have studied real analysis, you will be able to understand the proofs in the book. Complex numbers are surprising. Calculus with complex numbers turns out to be closely related to number theory and hyperbolic geometry. For this class, you need to know calculus.

This list is by no means exhaustive. University mathematics departments offer courses in differential equations, Fourier and Laplace transformations, geometry, cryptography, and so on. This list is just to get you started.

For students who already have a degree in another subject and would like to go to graduate school in mathematics, the most important courses to take are real analysis and modern algebra.

We have included this section because most students are not aware of the great variety of mathematics classes, and do not realize that there are many proof classes which do not require a knowledge of calculus.

For more information, we suggest that you talk with professors and advisors in your local university mathematics department.

# Dr. Spencer's Mantra for the Relief of Anxiety that Accompanies Attempts to Create and Write Proofs

The *Mantra* consists of three parts: the *Lines*, the *Refrain*, and the *Command*.

**The Lines.** To be chanted slowly as the situation demands. It is suggested that if a line contains $k$ words, then it should be chanted $k$ times, shifting the emphasis from word to word.

1. What am I trying to prove?
2. What are the hypotheses? (What may I assume?)
3. What is the conclusion I seek?
4. What do the elements look like?
5. Why do I believe it? (What do the examples show?)
6. What if it weren't so?

**The Refrain.** To be chanted between the lines, slowly and with feeling.

*What does that mean?*

Of course, the mood should be changed as necessary, i.e. sometimes the Refrain should read **What *should* that mean?**

**The Command.** To be followed regularly, after chants.

*Write it down.*
*Write it down carefully.*
*Write it down completely.*

*If symptoms persist, see your local professor.*

# Bibliography

Barnett, Stephen., *Discrete Mathematics: Numbers and Beyond*, Addison Wesley Longman Limited, Harlow, England, 1998. An easy-to-read and entertaining book about combinatorics, number tricks, and codes.

Burton, David M., *Elementary Number Theory*, McGraw-Hill, New York, 2002. A good textbook for a first course in elementary number theory.

Cantor, Georg, *Contributions to the Founding of the Theory of Transfinite Numbers*, translated and edited by Philip E.B. Jourdain, Dover Publications, Inc., New York, 1955 (reprint of 1915 edition). A short early treatise on transfinite arithmetic, almost identical with the modern theory.

Cateforis, Vasily C., *Math 547: Theory of Sets*, SUNY Potsdam, New York, 2002 (lecture notes, with exercises). An unpublished text for a course in advanced set theory and logic, based on the second half of *Set Theory with Applications* by Lin and Lin.

Epp, Susanna S., *Discrete Mathematics with Applications*, 2nd ed., Brooks/Cole Publishing Company, Pacific Grove, California, 1995. A textbook on proof, logic, and discrete math. Contains much good material but is too large, in our opinion, to be truly useful as a textbook.

Fraenkel, Abraham A., and Yehoshua Bar-Hillel, *Foundations of Set Theory*, North-Holland Publishing Company, Amsterdam, 1958. An entertaining history of early set theory, written by a mathematician and a logician.

Grimaldi, Ralph P., *Discrete and Combinatorial Mathematics: An Applied Introduction*, 3rd ed., Addison-Wesley Publishing Company, Inc., Reading, Massachusetts, 1994. A large book about combinatorics, graph theory, recursion relations, and so on. Good as a reference book, but too large for a textbook in a one-semester course.

Halmos, Paul R., *Naive Set Theory*, Springer-Verlag, New York, 1974. A very concise introduction to transfinite arithmetic, without many exercises.

Herstein, I.N., *Topics in Algebra*, John Wiley & Sons, Inc., New York, 1975. A very good textbook for a course in modern algebra.

Lay, Stephen R., *Analysis with an Introduction to Proof*, 2nd ed., Prentice Hall, Englewood Cliffs, New Jersey, 1986. An introduction to real analysis, written in a simple style and including some basic logic and set theory.

Lin, Shwu-Yeng T., and You-Feng Lin, *Set Theory with Applications*, Book Publishers, Inc., Tampa, Florida, 1985. A good, short textbook on logic and set theory including transfinite arithmetic but not much combinatorics. With the kind permissions of the Lins, we used the first half of their book (now out of print) as the foundation for our book.

Littlewood, J.E., *Littlewood's Miscellany*, edited by Béla Bollobás, Cambridge University Press, Cambridge, England, 1986. An amusing collection of reminiscences, anecdotes, mathematical jokes, and odds and ends, by one of the greatest mathematicians of the twentieth century.

Liu, C.L., *Elements of Discrete Mathematics*, 2nd ed., McGraw-Hill Book Company, New York, 1985. Another book sometimes used for an introductory course in set theory, logic, and discrete mathematics. This is a good book, concise and well written, on combinatorics, graph theory, recurrence relations, and so on. Liu's book is suitable for students who have already worked through our book.

Munkres, James R., *Topology: A First Course*, Prentice-Hall, Inc., Englewood Cliffs, New Jersey, 1975. A good textbook for a first course in point-set topology.

Rosen, Kenneth, *Discrete Mathematics and its Applications*, 4th ed., McGraw-Hill, New York, 1999. Another discrete mathematics textbook. Good as a reference book but too large for a textbook.

Ross, Sheldon, *A First Course in Probability*, 6th ed., Prentice Hall, Upper Saddle River, New Jersey, 2002. A good introductory probability textbook.

Rucker, Rudy, *Infinity and the Mind*, Princeton University Press, Princeton, New Jersey, 2005. An entertaining book about the history and paradoxes of set theory and the idea of infinity.

Rudin, Walter, *Principle of Mathematical Analysis*, 3rd ed., McGraw-Hill, Inc., 1976. A good textbook on real analysis, less elementary and more comprehensive than the book by Lay.

Vilenkin, N., *Stories about Sets*, Academic Press, New York, 1968. An entertaining popular book, originally published in Russia, about set theory and infinity.

# Index

# List of Symbols